THE JOHN DANZ LECTURES

THE JOHN DANZ LECTURES

Previously published in this series:

THE HUMAN CRISIS
By Julian Huxley

OF MEN AND GALAXIES
By Fred Hoyle

THE CHALLENGE OF SCIENCE

THE CHALLENGE OF SCIENCE

By George Boas

University of Washington Press SEATTLE

Library of Congress Catalog Card Number 65-23907
Printed in the United States of America

to James Baird

The John Danz Lectures

In October, 1961, Mr. John Danz, a Seattle pioneer, and his wife, Jessie Danz, made a substantial gift to the University of Washington to establish a perpetual fund to provide income to be used to bring to the University of Washington each year "... distinguished scholars of national and international reputation who have concerned themselves with the impact of science and philosophy on man's perception of a rational universe." The fund established by Mr. and Mrs. Danz is now known as the John Danz Fund, and the scholars brought to the University under its provisions are known as John Danz Lecturers or Professors.

Mr. Danz wisely left to the Board of Regents of the University of Washington the identification of the special fields in science, philosophy, and other disciplines in which lectureships may be established. His major concern and interest was that the fund would enable the University of Washington to bring to the campus some of the truly great scholars and thinkers of the world.

Mr. Danz authorized the Regents to expend a portion of the income from the fund to purchase special collections of books, documents, and other scholarly materials needed to reinforce the effectiveness of the extraordinary lectureships and professorships. The terms of the gift also provided for the publication and dissemination, when this seems appropriate, of the lectures given by the John Danz Lecturers.

Through this book, therefore, the fourth John Danz Lecturer speaks to the people and scholars of the world, as he has spoken to his audiences at the University of Washington and in the Pacific Northwest community.

Contents

Introduction

The previous John Danz lectures have been given by men responsible for delivering that impact on man's perception of a rational universe in which Mr. Danz was particularly interested. The present series is given by one who has been receiving that impact since his youth, some fifty years ago. Graduating from college on the eve of the First World War, he has seen the culture of the United States becoming steadily more technological. There is hardly an interest that has not been modified by applied science since that time, whether it is concerned with what used to be called the mind or with what is still called the body.

The three lectures that follow take as their starting point the idea that the scientific impact has been by way of a challenge. That challenge has come not so much through the conclusions of any science as through the scientific methods of discovery. Empirical investigations have taken the place of authority, and there is no longer any profit in deducing inferences from a priori principles. Thoughtful scientists, however, become philosophical when they reach the last pages of their books and often go in for speculative exercises that seem hazardous to a professional philosopher. They sometimes seem to forget that what we think of as Nature with an initial capital is a projection from the methods used to systematize one's data. These methods themselves are open to critical scrutiny and have been scrutinized by many philosophers. If scientific laws are statistical generalizations, as some scientists have said, then Nature itself is made up of collections of individuals. If on the other hand laws are universally and eternally true, then Nature turns into something lying beneath or above or behind the observable data. If they are the unavoidable ways of human thinking, as some have said the Law of Contradiction is, then the philosopher will insist upon the subjective element in all science. And if, as Newton said in the General Scholium to the *Principia,* they are edicts issued by a divine lawgiver, then Nature is what theology says it is. Such concerns usually do not worry the practicing scientist. He is quick to turn them over to his philosophical colleagues,

like a mother who tells her inquiring child that he will find out when he grows up. Unfortunately, although children do grow up, men merely die.

Every philosopher has certain presuppositions. I have myself assumed that things and events are interconnected in various ways and not in only one. Therefore I cannot believe that to unravel any one strand of interconnection as if it revealed the essential nature of things is a reliable method of knowing the truth. I realize at the same time that, unless one does single out the various strands, one cannot have scientific discourse. But to say that a material object is only what can be described in physics, rather than in chemistry or biology or even economics, is to eliminate from reality most of the things from which human problems arise. I have also taken it for granted that logic and the sciences can talk about individuals only in so far as the individuals in question are perfect specimens of classes. Pretty nearly everyone would make an analogous assumption. But I would differ from many of my colleagues in assuming also that although a single localized and dated perception, an individual experience, is not articulate and "says" so many things that it says nothing, it is the one unquestionable reality that we have. By all this jargon I mean, I suppose, that we are confronted with existent things and events and not with the intellectual constructs that are found in both science and philosophy. But here again I have to differ from those of my colleagues who make the same assumptions: this confrontation with existence does not nauseate me or arouse any anguish in me and, though existence is a logical surd, I see no point in punning and calling it absurd. The situation is that described by Morris Cohen in his *A Preface to Logic,* when he said, "It is as impossible to derive physical or psychological truth from pure logic as to build a house with nothing except the rules of architecture." This, I should imagine, could be accepted as the case without tumbling into lamentation.

I have also followed my custom of raising instead of solving problems. I have seen, I trust, the problems that reside in the obvious. It is a common failing of philosophers to believe that all their premises are either indubitable or self-evident, and I may have committed that error without being aware of it. Hence if any of my premises seem dubious, let them be accepted for the sake of argument, out of charity to a philosopher who has always been too skeptical for his own good. It seems obvious to me, for instance, that no one can expect an artist to derive from his work of art what a spectator derives from it. To paint a picture and to look at one are two radically different experiences. It has also seemed obvious to me that the theorems of science are all purified of their historical relevance; they are supposed to be true regardless of dates and places. Finally I assume that one can be religious without knowing anything of theology or ecclesiastical organization or history, and that theologians and church historians need not be religious. This does not mean that the religious experience, ecclesiastical history, and theology have no influences on one another, but the influence is not inevitable. Such premises as these have been disputed and will be disputed whenever they are pronounced. But then philosophy is always in a state of turmoil, for it is the only intellectual discipline that can belong to an individual. It would make no sense to ask a scientist, "What is your physics or chemistry or biology?" But it does make sense to ask a philosopher, "What is your philosophy?" This is either because philosophy has no settled bases or because there is a lyrical element in all philosophy. I assume the latter alternative. I have no more love for unity than for variety and think that truth often emerges from discussion. I do have a revulsion against authoritarianism whether it is found in religion, politics, art, or education. I have seen it in action in both wars and at first hand, and my resentment arises not merely because of the harm it does to others but because of that which it does to the man who wields it. The wielders of power are its most pathetic victims. This being so, I may lean too much toward variety. As a matter of fact I see no inherent value in either variety or unity.

But there is another factor involved here. Disagreement in philosophy demonstrates something important about knowledge. People on the whole disagree either because of their failure to use words in the same sense, because they accept different premises, or because of faulty inferences. The first and third of these causes might be eliminated, but I doubt that the second can. For in the choice of premises there enters a temperamental determinant. It may enter because of the emotional aura surrounding certain very abstract terms, like "unity," "eternality," "immutability," "regularity," "creativity," "progress," "vitality"—in short because of what A. O. Lovejoy called metaphysical pathos. Such terms name nothing that is given; they represent one of the goals of thinking. Moreover, we often accept premises because we are accustomed to them. Such premises become self-evident; they do not begin by being so. We often forget that we do not begin to do any systematized thinking until we are already formed by some educational process. We are seldom given the opportunity to dig out all the reasons that are habitually given for the process as we have it. We do not even state the conditions of our day-by-day perceptions of colors, sounds, smells, weights, lengths, and so on. We announce them as if they were unconditioned. Here even technical epistemology follows tradition. It seldom recognizes that every idea has a history and is not nontemporal. One can, of course, write out a set of ideas in propositional form and think only of their logical interrelations. I am not deprecating this practice except when its practitioners think they are talking about the origin and validity of our thoughts. But should anyone like to substitute the word "belief" where I say "knowledge," I have no objection. For some philosophers prefer to use the latter term for beliefs which they think are true. How they become true does not interest them.

One may still doubt, even after listening sympathetically to what I shall say, that I have clarified the challenge of science. Have I not rather simply shown the limits of science? The answer to this question depends upon one's theory of knowledge. If knowledge is exclusively scientific, then the challenge of science to art, philosophy, and religion is, "Go and do likewise." But I have tried to show in these lectures that this is impossible and why. If, on the other hand, one is willing to admit that the communication of experience, even in inadequate words or other symbols, is cognition, then it is art, philosophy, and religion that have issued a challenge to science. A further problem, and one which I have not dealt with in these lectures in any detail, is whether man's scientific or nonscientific explorations of nature are paramount. Let me say briefly and dogmatically that the answer to that question will be determined by one's temperament. For in the long run it will emerge from one's philosophic premises, and premises are a matter of faith.

These few paragraphs must suffice to clarify the point of view from which I have been writing. I should like to conclude by thanking the University of Washington for so generously allowing me the opportunity of putting my ideas before the public, my wife for listening to them and giving me the benefit of her comments about the first of the lectures, and my colleague, Dr. Harry Woolf, for reading the manuscript and helping me make it less amateurish than it would otherwise have been.

G.B.

February, 1965

Ruxton, Md.

THE CHALLENGE OF SCIENCE

The Challenge to the Arts

It is in no spirit of denigration that I point out that a science is always the record of someone's experience. There is nothing else that it could be. It is, of course, a record that has been purged, so to speak, of everything personal. It is thus not a lyric exercise. The techniques of purgation are well known and need only to be mentioned. To set up laboratory conditions is to erect a context in which general situations can be produced. It is a way of eliminating the things that have to be equal, as the phrase puts it, if the record is to be accepted as true. An experiment that is well conceived could presumably be carried out by anyone with the normal background of relevant information and sufficient manual skill. It is therefore sometimes thought of as having a kind of pure objectivity about it, the kind of objectivity that a camera might have. And yet, paradoxically enough, when a scientific law has been established in so perfect a fashion as to be clear and plausible to all, the name of its formulator is tacked on to it and it becomes Boyle's Law, Gresham's Law, and even Goedel's Theorem.

Now a moment's reflection will show that experience is not impersonal. Nor can one man's be possessed by another. Nothing is more exclusive than a given experience. Its date and the place where it occurs, for instance, are unique and may not merely furnish the context in which the experience happens, but also determine what it is like. There are many experiences that would be disagreeable at night but are pleasant in the morning, noisy ones for instance. But I shall resist the lecturer's temptation to list such factors as the contribution of distance to visual experience, of the medium through which sound waves travel to auditory, of the relation between taste and smell to gustatory, and of temporal succession to all. I shall also resist the temptation of listing all the personal factors that influence the nature of an experience, such as habit, anticipation, emotional states. Such things, if not already discovered, can be found in any elementary textbook of perceptual psychology. My introduction of such matters here is simply to suggest the difference between a scientific experience and an ordinary or day-by-day one. The former belongs to everyone, the latter to a single person.

If, then, I am not mistaken, it is wrong to speak of the sciences as experience in the same way that we speak of autobiography, lyric poetry, falling in love, or a religious conversion as experiences. And though there is sometimes a certain satisfaction in discovering that others have had the same kind of experience as oneself, there is also as much in discovering that no one has. It is, I imagine, like the satisfaction of a discoverer or inventor. To be accurate, no one can have the same experience that I have had, though he may have had the same kind of experience. In other words two men may fall in love, and it has been known to happen that two men have fallen in love with the same girl. But surely neither would say that the other's love was his love. To call the two experiences by the same name, "falling in love," is to rise above the level on which they occur to a logically higher level where the subjects of the experiences are eliminated. If Antony's love of Cleopatra was not identical with Caesar's, that would seem to be because Antony was not Caesar. And had the two Romans been in love with Cleopatra at the same time, their jealousy would not have been mitigated by the thought that love is a Platonic universal, the same no matter who is possessed by it. The self as the center of such experiences is what counts. For love is not an eleemosynary enterprise.

6 But that is true of all immediate experience. Whether one is involved in an emotional state, in a dream, even in plans for the future, the center from which the action spreads is oneself. And the goal toward which it tends is oneself too. Like the serpent which is depicted swallowing its tail, so here too the end is in the beginning. There is no doubt that such events can be described in general terms, but if expression is different from description, it would be impossible to express them in any terms other than individual. It is precisely at this point, as has been known for years, that scientific knowledge becomes inadequate. It is not inadequate for generalized descriptions, but it falls short when one hopes to communicate the "feel" of an experience. For that, one turns to art.[1]

[1] The difference between general and individual sentences is probably known to everyone, but they may be distinguished by their subjects. The subject of a general sentence is always a common noun, of an individual sentence a proper noun or its equivalent. The difference of which I have been speaking in the lecture follows from this. If you say that man is a rational animal, then of course every man will be a rational animal. This presents little in the way of difficulties. But suppose one has a definite man in mind, John Doe. Rational animals form a class of animal, and John Doe will belong to that class. Unfortunately, he belongs also to a number of other classes which in combination may not be attributable to any other man. He will come from a given family, with a certain heredity and environment; he will either belong to some church or not; he will

There are two sides at least to every aesthetic event, and two approaches to it, that of the artist and that of the spectator. This is but the most rudimentary account of the matter, but it will suffice for our present purposes. The artist himself, for reasons that vary from man to man, wishes to communicate, or at least to express, some experience or other. The tang of artistry to the artist must vary with the subjectivity

have a degree of formal education; he will have certain avocations—sports, music, reading, and so on—in combination or singly; he will have his own way of earning his living or will have private means; he will have his own habits and hobbies; he will be married or single or divorced or a widower, with or without children who will be many or few, predominantly girls or boys; he will have friends who will have some influence upon him; he may belong to one or more clubs, fraternal organizations, or societies with artistic, scientific, or political aims. Then his biography with all its successes and failures must be accounted for along with his reaction to them. All of these predicates are classifications whether they seem to be or not. To list them all would not be impossible, but it would be very difficult. Moreover, each man has his own way of being a son, brother, father, and husband; of carrying his education, his religion, his politics. He may be a faithful Catholic and a liberal in politics, conservative in his artistic tastes, generous in public charities, and hundreds of other things that have no logical interconnections. To find one word to describe him would naturally be impossible. One could only label him. And the label would be his name. This item of particularity or individuality was what Duns Scotus called the "thisness" of each particular, its *haecceitas,* as contrasted with its "whatness," its *quidditas*. The individuality of human beings has now been studied by scientists with very interesting results. See especially, Roger J. Williams, *Biochemical Individuality* (New York: John Wiley and Sons, 1956; also Science Editions paperback), and P. B. Medawar, *The Future of Man* (New York: Basic Books, 1960; also Mentor paperback). My own inferences from this situation can be found in *The Limits of Reason* (New York: Harper Bros., 1961).

of what is being expressed. It could not very well be otherwise. People too, even when they are not artists, vary in the extent to which they are self-confined. Some seem to succeed in almost eliminating their personal feelings, and all experiences become detached from their source and lie open to the public. A life of this kind, if pushed to its limit, would be like that of the perfect Stoic, properly called the life of apathy. In such a life there is no good or bad, no beauty or ugliness, no preferences: everything that happens is just another fact. Its one great difference from the life of the scientist is that it is simply a series of individual events and has no general relevance. Hence there is a legitimate question to be raised of why it should be said to be cognitive. For just as it does not contain any good or bad, so it does not contain any true or false. Its data are just momentary impressions, as Hume would have called them, and since all curiosity has been stifled and all passion spent, the data say nothing and raise no problems. A monster of this kind never existed of course, though he has been invoked in books on epistemology.

At the other extreme is an equally strange monster. He is the person who finds it impossible ever to envision things objectively. He believes that everything that happens, happens to him. He is therefore subject to the pathetic fallacy and reads emotional expressions in every landscape. He easily turns into the kind of man who sees a supernatural being in every tree, every stone, every shower of rain, to say nothing of some local god who either loves or hates him. We all probably indulge in this sort of self-gratification from time to time, for the very business of living demands that we assume both attitudes. For if we are too objective, then nothing matters, and if we are too subjective, then everything matters, and there is no scale of importance. But what is the meaning of "too" in these clauses? That unfortunately can be defined only in the context of a general philosophy of life. And one soon finds oneself arguing in a circle. We just assume that some things and acts are more important than others, in which case we have to appraise objectivity in relation to our scale of values. Or we assume that nothing is more important than anything else and conclude that we cannot be too objective. May I say that when I have read the works of philosophers of this type, I have found that they are just as hot-headed in fighting for apathy as the most sensitive poet in fighting for his right to self-expression.

From the spectator's point of view the work of art is first of all a physical object, located out there in time and space. But he also postulates that it exists as a message to him. In so far as he claims to be nothing more than a specimen of the human species, he may be right. No one is as colorless as that, but individuals do differ so much in their self-assertion that there may be human beings who are uninterested in being anything more than that. Their search for nonexistence will probably never be fulfilled short of suicide, but their program can still be understood. The trouble is that they are not only content in their humility, but often speak as if their fellow men ought to follow their example. They want to turn into an item of the universal *Mankind,* and they assert that only those messages which are intelligible to all men are worth emitting. But, as in all human affairs, there are other individuals who either will not or cannot take this point of view. They know that the artist does not even know of their existence and that the interpretation of the work of art which they are viewing is their own and nothing more or less than their own. It is they who recreate the work of art, absorb it into their total experience, give it what is called a meaning that may not have been that of the artist at all. The impressionistic critic is perhaps the best example of such people. Their attitude is about as antiscientific as a report on experience could well be.

The greatness of science, as we all know, lies in its laws. The individual's personal reaction to a work of art is the very antithesis of law. The psychologist, to the extent that he is a scientist, will quite rightly eliminate the personal equation in the aesthetic experience and attempt the construction of laws that will describe it in terms of an overindividual, generalized human life. He will refer to the human being who is supposed to live this life as "One," "We," and sometimes "You." He will write sentences like these: "One feels, standing before the 'Mona Lisa,' the enigmatic character of womanhood"; "We look at the Parthenon and feel the glory that was Greece gleaming from its tawny marbles"; "You can't read *Mme Bovary* without a shudder at the poverty of provincial life." Such sentences are of course patently absurd. Plenty of people have stood before the "Mona Lisa," looked at the Parthenon, read *Mme Bovary,* and been bored to death. Yet to cast one's own impressions into the mold of everyone's is to look scientific when one is really being egocentric and so full of one's own importance that one inflates one's ideas into universals. This is not intended to imply the impossibility of generalizing about either the source of works of art in the human psyche or their effects on the public. It simply repeats and emphasizes the truism that an individual's experience is individual and may not be typical of anything. The challenge of science to both the artist and the spectator or critic is found precisely here. It can be stated in the following questions:

1. What is to be substituted for the universal and everlasting values of scientific truth?

2. How is one to verify an expression of a purely personal experience?

3. What justification is there for such self-indulgence as is found in lyrical expressions?

4. Will not lyricism inevitably lead over to unintelligibility?

These questions are legitimate and have been frequently asked. And it must be acknowledged that the answers I shall give to them would be unacceptable to most of my colleagues. But that consequence will have to be accepted. Yet, since we are talking about the impact of the sciences upon the arts, I am maintaining that their greatest impact comes from the very nature of verified generalizations, not from any single theorem or set of theorems such as any particular science might demonstrate. Art will not necessarily change because the universe is expanding or the population increasing too rapidly for comfort. But it might change if scientific method were to shift from empirical proof to intuition.

First, then, the value of the universal.

The emotional aura of the terms "universal" and "eternal" is very satisfying, and it might be maintained that only those works of art which everyone everywhere finds beautiful are worthwhile. But there are certain objections to this point of view. If it is the universal attraction of a thing that makes it great, then one ought to be able to survey all the things that are universally liked and see how great they are. But as soon as one begins this survey, one has to include obscenity, which is found everywhere among all peoples and in about the same form. Anyone who has lived in intimacy with large groups of his fellow men, as in the armed services, knows pretty well that the obscene joke is a favorite whereas the witticisms of a La Rochefoucauld or George Meredith fall flat. Contemporary writers in English seem to understand this and are doing their best to restore obscenity to the realm of the admirable. There are also certain human attitudes of which we are supposed to disapprove, such as hypocrisy and cruelty, and they are much more widespread than the virtues antithetical to them, sincerity and kindness. Surely theologians of the Augustinian persuasion would not have taken such pains to elaborate the dogma of congenital wickedness unless they believed evil to be more general than good. Reflecting on such data as these, one is bound to come to the conclusion that ethical goodness is much rarer than badness, that the masterpieces in art are fewer than the failures, and that truth in science is less frequently discovered than the plausible which turns out to be wrong. In fact most people, I suspect, would say that this was obvious. But if it is obvious, or if it is simply true, then what is universal about the value of art? Even when a given work of art has been admired over a long period of years, as the masterpieces have been admired, one finds that the admiration is based on varying grounds, now

being naturalistic realism, now the moral truth of the subject, now a kind of idealistic subject matter, now the originality of the artist, and so on. One cannot logically hold that the universal is better than the particular and also hold that only a few masterpieces, saints, and seers exist.

In fact I am willing to advance the thesis that when one loves or admires something deeply, one cares very little whether anyone else does or not. One is more likely to drag in the others when one is in a mood of self-justification. The child when reprimanded says, "But all the fellows do it. . . ." And the man, "I must not be eccentric."[2] There have been few, if any, great artists who gave a thought to what society thought of them or of their works, though they naturally enough enjoyed the approval of others when it came to them.

[2] One is in a curious position here. On the one hand we are told that the good is the universal, and this is backed up by such phrases —of reprimand—as "that is not done," at the moment when one is doing something which is believed to be wrong. On the other hand the bulk of the books on goodness and beauty is not given over to an exposition of the most usual traits of action and of art but, on the contrary, to the least usual. If you pursue the line of universality, then ethics turns into a study of manners, *les moeurs*, mores, custom. And though one has a compulsion to fall in with what "everyone" is doing, one is criticized for doing so. The burden of books on ethics and normative aesthetics is to make behavior uniform, whether in matters of daily life or in matters of taste. But this would be unnecessary if norms were really universal.

Was Beethoven a conformist? Was Michelangelo one to knuckle under to popular opinion? I am not saying that great art cannot be produced by submissive souls, for it clearly can. It must be granted that social approval is doubtless very pleasant, and some people will go to any lengths to attain it. But there are moments when self-assertion seems more valuable. My point is merely that universality is irrelevant to value.

The cult of originality actually arose in the nineteenth century. It was no longer important to be like anyone else, to say nothing of being like everyone else. The advice of Polonius to his son was taken seriously. The trouble was, I grant, that few of those who talked the loudest about being true to themselves had any selves to be true to. The antics of some of the early Romanticists, like those of some of our contemporary Beatniks, were a mode of creating a self rather than of expressing one. Nevertheless one cannot but feel that the value of self-expression up to the point at which it becomes self-destruction is probably very great. This would seem to be particularly so in a society where the interconnections among its members are binding. There would, for instance, be little sense in expressing one's individuality in a monastic order, in the United States Marine Corps, or in a typing pool. In such societies rules are more important than self-expression, and it is taken for granted that the rules will be obeyed. There have been writers like Gertrude Stein and James Joyce, who have violated almost all the rules. But that is not what made them important artists, if they are important artists, though it did capture the fancy of their readers. But as in trying to balance a checkbook we submit our wills to the rules of arithmetic, so in writing a sentence we are more likely to submit to the rules of grammar and syntax than to make up rules of our own. That is if we know what they are. But no one since the seventeenth century ever thought that obedience to the rules of art, unsupported by something more, would make an artist better than the run of the mine. Indeed I suspect that the reason why admiration for correctness survived as long as it did is that it provided critics with a measure by means of which they could tell whether a work of art was good or bad. And that was their livelihood. It is easy to tell whether a draftsman has obeyed the laws of visual perspective, but it is harder to appraise intensity of expression.

It is interesting, if not more, to observe that the cult of individuality spread at about the time of the Industrial Revolution. The unevenness of handwork, which has often been admired, would be fatal in machine work. It would prove that the machine was incapable of doing a perfect job. The beauty of machine work is precisely its uniformity, and where the machine is needed originality is out of place. We want our power looms, our adding machines, and our measuring instruments never to deviate from the fraction of tolerance that their inventors—or their employers—have set up. An original typewriter that could not be counted upon to print the letters attached to its keys would normally be sent back to the factory. The famous hundred monkeys would be poor competitors with such a machine. Moreover, the operators of the machines are required to be as regular as the machines they operate. Their skill is appraised by their success in subordinating themselves to the rules. The demand for originality, which sometimes went to extravagant extremes, was an effective balance to this kind of life, though there was more fear of the machine than was rational in this demand. Yet during the last century and a half the greatest emphasis has been put on the necessity for individualism, originality, self-expression. And just as production has been made more and more automatic, so nonconformism has been more and more preached as a way of life. In short, universality in artistry was more plausible as a criterion of value when the organization of living was shaky; originality or individualism became the standard of value when it became firmer.

We have now reached the point where no work of art can be understood according to any generally accepted code of interpretation. I am neither deploring, nor for that matter extolling, this. At the present in New York, London, Paris, or even Tokyo, one can see exhibitions of paintings in which the pictures represent every aesthetic fashion that has appeared on the scene since 1800: photographic realism, impressionism, abstractionism, actionism, tachism, and no doubt half a dozen other doctrines. The focus swings from the purely objective in which the personality of the painter is suppressed, as far as that is possible, to the purely subjective in which one would have to identify oneself with the painter to appreciate what he has been trying to do. One's appreciation of these forms of painting will vary, but I imagine that one's tolerance for the most hermetic will depend upon one's submission to authority, which in turn will depend upon one's contentment with regularity or habit.

One of the scientific sources of the acceptance of originality is Freudian psychology reinforced with recent advances in biochemistry. If the psychic differences among individuals are congenital, and if they can be linked to biochemical variations, then there is nothing that the admirer of the universal can do about it. He will just have to accept the hard fact that people differ not only anatomically and physiologically, but also ethically. It may very well be true, for instance, that recalcitrancy and submission are built into the human being at birth, though no one probably is recalcitrant to everything or submissive to everything. Life is just not like that. To live is to meet problems and solve them, and there is no way of predicting just what problems any individual will have to meet. Nor is there any way of predicting just what kind of problem any individual will be able to meet. This was pretty well shown during the Second World War in the Underground, where people of all social stations, of all economic levels, of all religious and antireligious backgrounds, of university education and of lower schooling, worked together to defeat the Nazis—and this in the face of powerful police forces and almost certain torture and execution. We can say without fear of contradiction that a man with only one arm will not become a violinist and that a low-grade moron will not become an astronomer. But that is about as far as such predictions can go. One has only to study the early work of painters and writers to see how hard it would have been to predict their mature productions. The young Keats, the young Tennyson, and even the young Milton wrote trivialities.[3] The paintings by the young Rodin gave no

[3] This may seem exaggerated. But read Keats's *To Some Ladies,* the opening verses of which are:

> What though while the wonders of nature exploring,
> I cannot your light, mazy footsteps attend . . .

As for Tennyson, his follies did not cease with his youth. Read his *Ode Sung at the Opening of the International Exhibition.* Of Milton's

evidence of the sculpture that he was to create at a later date. The young Renoir was a china painter. The young Daumier was a commonplace caricaturist. When Leonardo was working in Verrocchio's studio, he painted like Verrocchio, and when Raphael was under the orders of Perugino, he painted like Perugino. The study of individual differences has shown us that individuality is ineradicable, though a man may be beaten into submission or driven to abandoning his program of life.

All I have done so far is to suggest that perhaps the universal is not so reliable a criterion of value as the books say, and that life in an age like ours demands originality to counterbalance imposed conformity.

The second challenge which a scientist has every right to put to an artist is how one can verify the expression of a personal experience. The artist cannot reply that anyone can do this for himself unless that which makes his work of art an individual experience has been eliminated. Strictly speaking, no unique experience can even be expressed, to say nothing of being verified. No man can feel another's pain, and there is

---

early poems, William Vaughn Moody wrote, "It is hardly wrong to say that the English poems which Milton wrote before his twenty-third year are interesting chiefly because of their defects"; and again, "He was surprisingly tardy in finding his voice." See his edition of *The Complete Poetical Works of John Milton* (Boston and New York: Houghton Mifflin, 1924), p. 5.

a degree of fortitude that will keep pain unexpressed. If the victim wishes to conceal his sufferings, he can do so. And the long-faced boor may also conceal his pleasure. But though we all know that we cannot communicate our feelings satisfactorily, we can approximate to a satisfactory communication of them. On the other hand, a lyric poem, a painting, a musical composition—these things are not simple declarative sentences. As Milton said of books, they are not absolutely dead things but can be as active as that soul whose progeny they are. The action that they produce is somewhat mysterious, for it seems to consist as much in stimulating a kind of sympathy in another person as in expressing one's own feelings. The result seems to be that two people participate in the same experience. To put this into words, however, is to translate something nonverbal into an inadequate code. To translate the poetry of one language into another is always to create a new poem based upon the original, or to communicate the intellectual content of the original and nothing else. There is a poem of Verlaine that begins:

> Je fais souvent ce rêve étrange et pénétrant
> D'une femme inconnue que j'aime et qui m'aime. . . .

To put this into English is to come out with:

> Often I have this dream, strange and penetrating
> Of an unknown woman whom I love and who loves me. . . .

If anything could be more of a treason than this, I would be hard put to it to find it. Or take the famous lines of Baudelaire:

> Là tout n'est qu'ordre et beauté,
> Luxe, calme, et volupté. . . .

I am ashamed even to experiment on this, for anyone who knows any French at all knows how the poetry of such lines evaporates when translated. To translate English lyrics into French is equally absurd, and even the simple lyrics of Heine resist translation into English. But this is a commonplace. The difficulty clearly is not merely that of verbal associations or connotations, rhyme schemes and metrics, but also the very intimacy of the poems. The problem would be easier to solve if a word were simply the name of something. But of course it bathes in an emotional cloud, and in literature the emotional cloud counts for a lot more than denotation does. Even if we had a complete catalogue of all possible emotional states and a lexicon of connotations, that would be no substitute for the experience of reading a given poem. It would be rather like chasing footnotes when reading Aeschylus and thinking that one is reading him as a fifth-century Athenian read him.

In short, the demand for verification cannot be met if we think that works of art are always assertive or declarative. We are in the same situation with regard to pictures and sculptures, to say nothing of music. One has only to read those program notes of concerts that attempt to translate music into words to realize how futile an exercise it is. Suppose that one is told that Beethoven's *Symphony in C-minor* means fate knocking at the door. The Philistine can always say, and with justice, what of it? The symphony is sounds arranged in a certain order and has stirred very deeply human beings who never heard of fate or who are unable to determine what fate knocking at the door actually means. In fact, the opening theme of that symphony has come to mean something quite different since the Second World War, and the generation that had direct experience of that horror will probably always think of it as either a shout of victory or a demand for it. What then happens to the other three movements I shall not attempt to say, unless the andante has come to mean the disillusionment that follows victory.

The proper answer of the artist or critic to the demand for verification would seem to be, "Verification applies only to declarative sentences." When a man shakes another's hand, the gesture cannot be verified: it is simply an expression of affection or of obedience to customary etiquette. It cannot be said to assert anything definite, but it can be translated into as great a variety of sentences as one chooses. But when one comes to the sixty self-portraits by Rembrandt[4] or Picasso's

[4] The number of self-portraits of Rembrandt varies according to the authority one uses. Adolf Rosenberg in his *Rembrandt des Meisters Gemälde,* "Klassiker der Kunst" (Stuttgart: Deutsche Verlags-Anstalt,

"Guernica" or Piero della Francesca's "Resurrection," how is he to put these objects into sentences? I doubt whether it makes any sense to attribute meaning to them in the ordinary sense of the term. They are a compression of one man's impression of a person or a historical event or a theological dogma, and they stimulate in the minds of some observers—though not of all—certain emotions that arise in part because of the observer's total experience. And that includes his education. The effectiveness of such a transfer of experience varies with the similarity between artist and observer. But there happen to be some works of art which are so powerful that one takes for granted the identity of one's experience with that of the artist on seeing, hearing, or reading them. Thus, paradoxically, the so-called message of the work of art is assimilated to that of a scientific theorem.

Now, the value of scientific theorems lies in their power to describe a world that lies well beyond the experience of any one individual. Nevertheless this world absorbs the experience of all men. To know that man is a rational animal, or

---

1906), II, 432-33, lists fifty-eight; A. Bredius, *The Paintings of Rembrandt* (New York: Oxford University Press, n.d.), plates 1-62, lists and reproduces sixty-two; Jakob Rosenberg in *Rembrandt* (Cambridge, Mass.: Harvard University Press, 1948), p. 23, writes, "We know altogether about sixty painted self-portraits by the master, in addition to more than twenty etchings and about ten drawings." It is perhaps unnecessary to add that the reason for these differences is in various opinions about the genuineness of some of the portraits.

the one animal that laughs, or the tool-using animal, or the one animal with a sense of sin, or the being that fits any of the other popular and profoundly misleading descriptions, is not to understand the man who stands before you, who tells you his troubles or confides his joys, who asks your advice, who grows angry at some real or imaginary wrong, who flatters or insults you (if the distinction is reasonable), who tries to persuade you to do something you do not want to do, who scolds you for something he thinks you have done. Such generalized descriptions, assuming that they are true, do help you to understand the species *Homo sapiens,* but not John Doe or Richard Roe. What is wanted is an intimate apprehension of the fellow's private experience, and, unless I am completely wide of the mark, only art will do that for you. To see the face that Rembrandt thought was his, to see it as he saw it, is to know Rembrandt even if you know little about him. And similarly to see Piero's "Resurrection" is not to read a treatise on Christian theology but to see the risen Christ as a man who has been for three days in hell, risen from the battle with death, expiation, self-sacrifice. How trivial seem all these phrases, however, when one is face to face with this fresco! What is more, they are futile anyway, for what is needed to understand the painting is not description or verbal paraphrase but the vision of the painting.

The impact of the natural sciences on the arts has switched them from their normal course toward a goal that they are by their nature incapable of reaching. The task they can perform admirably is promoting the sympathetic understanding of another's mind, but, when they attempt to be verifiable, they become pretentious and usually are based on false premises. Think of the realistic novels of Zola. His theory of genetics was wrong, and yet he thought his Rougon-Macquart series revealed an important scientific truth. The one thing that makes them readable today is their nonscientific parts, those parts in which Zola forgot his theories and instead vivified detailed events of character and the basic human emotions. But one does not have to return to Zola to illustrate this. Picasso maintained at one time that since a physical object has many sides and not merely that which faces the painter, all sides must be put down on the canvas at once.[5] To know this may help you to understand why Picasso at one time painted a conglomeration of aspects simultaneously, but it does not help you to be moved or even amused by what you are looking at. I am far from depreciating the pleasure that one can derive from

[5] This has no pretension of being the heart of Picasso's aesthetics. For a study of the ideas behind cubism see Christopher Gray, *Cubist Aesthetic Theories* (Baltimore, Md.: Johns Hopkins Press, 1953), especially chapters iv and v. See also Charles E. Gauss, *The Aesthetic Theories of French Artists* (Baltimore, Md.: Johns Hopkins Press, 1950), chapter vi.

virtuosity, from seeing how an artist has succeeded in doing what his program demands of him. There is always pleasure in observing a skillful performance, regardless of the intrinsic interest in what is performed. But that is only one part of the value to be gained from aesthetic experiences. I doubt, for instance, that anyone ever enjoyed the sound of one of Lizst's operatic transcriptions, but a pianist certainly can enjoy hearing another pianist overcome the technical difficulties of playing them. Similarly, when Renaissance painters became infatuated with perspective, they produced paintings in which very difficult stunts were carried out successfully. But that is a different matter from producing paintings that carry one into the heart of another man's life.

Let us suppose that what I am saying is untrue and that there is no way of producing this identity of experience between artist and observer. At least the illusion that we share another's feelings is a powerful one and one that has often alleviated the gloom and the suffering of life. The serenity of Mozart's *Symphony in G-minor,* the penetration of Renoir's portraits of children, the tragic accents of Bach's *Saint Matthew Passion* may all be illusions, and moreover illusions that are not universally shared. What of it? A death in the family may be of no moment to anyone but the immediate relatives of the deceased, and the cars continue to roar by the front door, the children shout on their way to school, the radio goes on croaking its commercials as if nothing had happened. Grief is no less grief because it has not spread in waves throughout society. A crimson sunset is no less crimson because it lasts for only a few minutes. In fact it might be reasonably maintained that the impact of science on the arts has been to make aestheticians remove the focus of art from persons to some celestial realm where no one cares about it at all.

We have raised the question of what justification there is for such self-indulgence. There would be justification enough if the enjoyment of works of art simply restored the individual to his position of being the center of experience. Our education as it has developed has left to the individual only a spot beside his test tubes and microscopes. There are universities making a courageous struggle to increase the individual's participation in the discovery of truth, but there is so much information to be absorbed before any creative work can be begun that he soon loses whatever impetus he may have started with. The premium put upon information is so great that the undergraduate has little time and energy to do any thinking on his own. I have no statistics on the productivity of men who have taken courses in creative writing, painting, musical composition as compared with those who have simply sat down to write, paint, or compose, but I suspect that those

Americans who have received Nobel Prizes in literature were not the products of courses in writing. This is not to say that the students in such courses have less innate talent than Sinclair Lewis, Faulkner, Pearl Buck, or Steinbeck, but merely that the value of a workshop is that of a springboard. No good ever came out of bouncing forever and not taking off.

But even outside of college the business of living sets up barriers to self-expression, and, what is worse, the time left for leisure is filled by organized entertainment over which the individual has little control. I realize that you can always turn a button, but you cannot choose what is to be the effect. Just as history is what the textbooks say it is, and philosophy what the textbooks say that is, and so on with all the other humanistic interests, so music is what comes out of the radio and painting what the museum is showing. This is inevitable, and one need not be a snob. But one cannot help wondering whether playing a quartet is not more important to a person than listening to one, and whether looking at the landscape is not better than listening to a lecture about it. For unfortunately the books tell us what to listen for and what to look for, and we are docile enough, or perhaps impressionable enough, to obey.

This would not be very serious if we did not transfer the same attitude to our personal relations. If it is the style to classify people as anal-erotic, or extrovert, or hyperkinetic, or other-directed, or whatever the current slang makes fashionable, we will treat our friends according to the rules given in the books we have read. But no one is always anything. And the same is true of works of art. What a man is in part—I do not say as a whole—is determined by his reactions to the person he is dealing with. The hyperkinetic man may be trying to get away from a bore and would linger near someone more interesting. Similarly, what a work of art is, is in part determined by the person looking at it at a given time and under given conditions. This is bewildering, but at least it liberates the spectator from the dictates of the arbiters of taste so that he can make up his own mind, assuming that he has any mind left to make up. The arts, both from the artist's point of view and from that of the spectator, are liberating forces, and, if one is not too high-minded to exclude humor, they join with laughter in lifting men out of the network of compulsions. There is nothing to laugh at in science, and one of the effects of applied science, if not of pure, is the notion that life is a grim, sordid affair in which there is no hope, and society is at best an anthill or a beehive. The inference is invalid but nevertheless widely drawn. And unhappily it has been drawn by artists as well as by philosophers. It is almost impossible today to see a painting that is free and exuberant. On the contrary, at the one extreme we see the results of the artist's throwing his colors about in a frenzy and calling the result action-painting, or at the other extreme planes, lines, and other simple geometrical figures overlapping and intersecting as in a kaleidoscope. No one would have the courage today to paint like Renoir or Pissarro. This does not imply that contemporary painting is bad, but simply that it has renounced human life.

It has often been observed that the arts have become more and more inaccessible to direct experience. One has to know first, as the phrase puts it, what the artist was trying to do. A manifesto lies behind each painting, sculpture, poem, and musical composition. One has to have a key. The work of art becomes a cryptogram. To decode it is a fascinating diversion for the student of such things, but it is mortally tiresome to the amateur who wants to sit down and just look, listen, or read. The paintings of the High Renaissance were just as complex as those of today, but they were usually not a puzzle. The amateur could get a lot of pleasure out of just looking at them, whether or not he understood the symbolism, the iconographical details, or the position in history of what he was looking at. But it was recently pointed out to me at an exhibition that the onlookers were walking about with brows knitted and eyes glazed; each picture was a challenge. This is largely true because painters have read a little about some science and have tried to incorporate it in visual terms. It is as if they had heard of the Heisenberg Principle and then decided that they must illustrate the uncertainty of both velocity and position by making masses interpenetrate and having splashes of color dart from corner to corner. Analogously the Surrealists tried to put what Freud said about dreams on canvas and to make their forms change by visual metamorphosis as one looked. But when the question was put to them of why they felt impelled to do these tricks, the reply was a kind of pseudo science that was unintelligible.

It is not essential that a poet or painter reproduce in his works the realm of nonscientific experience, but it is at least legitimate for him to do so. That is, after all, the realm in which we live, the realm of colors, sounds, textures, molar objects, sticks and stones and flowers and trees and human beings. It is not the realm of scientific objects, of electrons and protons and all the inhabitants of the nuclear kingdom. I mean that the problems that confront a man in his daily life are not those of quanta, of DNA, of the incidence of prime numbers, or of the expanding universe, wonderful as are the solutions to such problems. It is not to deprecate science that one says this. It is simply to indicate the level on which we live. To a man who puts salt on his breakfast eggs, it matters very little that sodium and chlorine are its constituents. When he pours water on a fire, he cannot be worried about the fact that water is composed of a highly inflammable gas and another which is necessary for combustion. When the impact of science on daily life makes men forget the level on which their problems arise, then indeed one could deprecate its study. When a child dies, its parents do not want to be told that death is nothing but the cessation of the pulse. They are not interested at such a time in physiology.

Our final question is whether indulgence in the transmittal of immediate experience is not bound to terminate in unintelligibility. In a certain sense, yes. For everything that is individual is unintelligible. The problem is rather whether an artist ought to confine himself to those experiences which are common to the majority or express whatever interests him. If he chooses the former alternative, he will run the risk of propagating platitudes and achieving the popularity of all who deal in platitudes. If he chooses the latter, he runs the equally great risk of obscurity. We know enough of the history of taste to realize both dangers, and we also know that some of the artists who have the most enduring fame were not understood when they were most active. In every work of art there is an area that lies open only to the man who made it. This is of course true of each man's life as well, whether he is an artist or not. It should be added that it is not true that all great artists failed to be appreciated in their time. Michel-

angelo and Raphael were highly praised early in their careers; so were Mozart and Haydn. Yet no one really knows any of these men. Each had a life that has remained a secret. We cannot know what any of them intended to do in any of their works of art, and the best we can do is to substitute our impressions for their intentions. Yet it is perhaps this hidden area which interests us more than that which lies open to our inspection and which tantalizes the critics of every age. There are, to be sure, some critics who prefer the easily understood, the commonplace, to the mysterious and recondite. They justify their preference on the ground that it helps continue a great tradition. So do the turtle and the lungfish.

However annoying it is to come upon something one cannot understand, there is always the possibility that someone else will understand it. And he may be able to transmit his understanding to his fellow men. One cannot flatly condemn as bad, works of art that are opaque or puzzling or meaningless. But at the same time one can condemn them as food for oneself. This does not seem to me to be true of science. One has the feeling that one ought to make every effort to inform oneself of what the sciences are up to. But that may be because no scientific conclusion exists exclusively for its framer. Each is supposed to have general reference. In the arts, however, the fact is that any picture or poem or what-not might cease to exist, and all that could be said is that the pleasure of living had been diminished. Yet everyone would know that other works of art were there to be enjoyed, and that in the long run no one was any the worse off. There must be millions of people who have neither seen nor will see Angkor Wat, have never heard or will hear the "Goldberg Variations." This has not made life less endurable, for works of art do not have to be known. But there is no one living today who, whether he realizes it or not, is not affected by the latest discoveries in nuclear physics. Science is by its very nature a social enterprise; art an individual one.

My point in this lecture, then, has been simple. It is that the impact of science upon the arts has been that of a basic challenge. Science has succeeded in establishing new truths in all its fields and has done so by reducing the personal equation to zero as a limit. The arts, if I am right, are engaged upon a task that is antithetically opposed to this. It is the transfer of the immediate experience from the artist to the spectator. There is, it goes without saying, an art of scientific exposition, and a great one too. But even here the excellence of the artistry, as opposed to the content of the communication, varies with the expounder, and of two scientists saying the same thing, one may be clear and the other confused, one convincing and the other implausible. But we are forced to deal with what happens on the whole, not with individual variations. Where the scientist is limited by logic, by his laboratory technique, by mathematics and the calculus of probability, the artist is limited only by sincerity. There is an ethics of artistry which is usually taken for granted, observance of which can never be proved. And, since it is taken for granted, we have the feeling when we are in the presence of a work of art that we see through it into the character of a man. If we are right, then the knowledge that comes from art is certainly as important as that which comes from astronomy, physics, chemistry, or biology. But it is useless knowledge unless mediated by philosophic reflection. We shall discuss that in our second lecture.

The Challenge to Philosophy

Philosophy and science were indistinguishable up to the end of the eighteenth century when Kant convinced the majority of philosophers that science could deal only with the world of appearance whereas philosophy dealt with the world of something called reality. There was never a more superficial divorce, for both partners kept reuniting and separating like so many couples who can neither live together nor live apart. Since the publication of the *Critique of Pure Reason* no two philosophers have been able to decide what their discipline should be. Some have distinguished it from other disciplines by its subject matter, others by its method. Some have claimed to be able to discover formulas that will unite all possible problems and solutions, others on the contrary insist that philosophy has an area of its own which no science has been interested in studying. William James said it was those problems left unsolved at any given period, problems which when solved became the province of a science. And indeed in my own youth there was no separate department of psychology, but that field was united with the field of the philosophers. (In spite of this Wundt had established his laboratory fifty-odd years before, and psychologists had begun fruitful investigations into perception and allied areas.) Now there must be few American universities where the two departments are not separated. But even when there is a separate department of philosophy, its occupants have abandoned some of the traditional problems, sometimes on the ground that they are meaningless. The existence of God, freedom of the will, and the immortality of the soul are still debated in certain places, whereas in others these subjects are merely embarrassing. Some philosophers say that they have no subject matter at all, but that they practice a peculiar method, usually that of linguistic analysis, and that any problems whatsoever can be treated philosophically. In Catholic institutions the philosophers operate on the principle of authority and try to work out the implications of Thomism in the light of contemporary issues. Others argue only about the grounds upon which any idea rests and spend their time trying to find a satisfactory basis for our premises. These, my philosophic colleagues will admit, are only a few of the many descriptions of the philosophic adventure, and they should all be supplemented by pointing out that there is also the idea of a philosophy *of* everything—a philosophy of science, of religion, of history, of art, of politics, and so on.

This being so, it is foolish for any man to pretend to know what philosophy is by nature, and I shall not indulge in that game. For the names that we give to the various learned disciplines are not given supernaturally or, if you prefer, *in rerum natura,* but are the tags and labels which remain on our intellectual luggage long after the initial voyage has ended. Any traveler knows how some people leave these labels on their bags as a testimony to their wanderings, and that would be a suitable metaphor to indicate where philosophy has been during its twenty-five-hundred-year-old history in the Occident alone. The name, which, as every freshman knows, means the love of wisdom, is after all not bad. For wisdom like stupidity may be looked for in any human endeavor, and I should imagine that a scientist or artist or historian could love wisdom just as fondly as a symbolic logician or linguistic analyst. It turns up whenever there is revealed a scrupulous criticism of the grounds of one's beliefs. The physicist becomes philosophical when he asks such questions as, "What is meant by a definition?" or, "What is the role of basic metaphors in explanation?" or, "How far can one eliminate the personal equation in knowledge?" But similarly the artist can ask, "What is the purpose of artistry?" or, "What are the limits of aesthetic communication?" or, "What is the relation of art to morality?" It may be that such questions can be answered only by fiat. We can leave that question alone here, though eventually some philosopher, if his devotion to wisdom be complete, will have to ask it.

As one surveys the history of philosophy, one sees that there has always been a give and take among philosophy, common sense, and the sciences. Philosophy has inherited problems from common sense, and it has amended the procedures of science. It has taken over from common sense, to give but one example, the illusions of ordinary perception, such as the visual convergence of parallel lines, and distinguished carefully between appearance and reality. It has taken over from physics a conception of a permanent and indestructible material world and inferred religious conclusions from the existence of that world. It has given to science certain rules, such as, "Nothing is made from nothing"; "Nature always follows the simplest course"; "Nature does nothing in vain"; and these and similar rules have determined the kind of scientific conclusions that would be acceptable.

One may have one's doubts about the value of some of these rules, to say nothing of their binding power, but the fact remains that men surely as intelligent as we are, obeyed them. Each of them is based on some metaphysical dogma, some dogma that is usually unexamined. It is unexamined because it had been part and parcel of collective thinking and seems self-evident.

The reason for the interplay of these three pursuits—philosophy, common sense, and science—does not seem to me to be very recondite. In common sense, though tradition and the exigencies of language play a greater part than thinking does, nevertheless we do not try to make solid objects interpenetrate, nor do we put a kettle of water into the refrigerator when we want it to boil. We time things, we measure things, we weigh things, and we do all this without first reading a book on the nature of time, on the effect of velocity on length, or on the deviations in weight due to altitude and latitude. Sometimes we find linguistics and common sense influencing the assumptions of even the greatest philosophers. Descartes, for instance, in spite of his claims to doubt everything, could not bring himself to doubt the need for a subject for every action. He took that over, presumably, from French grammar. In Osiander's introduction to Copernicus' major work, he lays down the dogma that nature always follows the simplest course, and presumably it is simpler to move small bodies around big ones than big ones around small ones. Common sense assumes that every instrument must have an end and hence interprets anatomy in teleological terms. But it is very difficult, if possible, to read a book on the anatomy of birds, beasts, and fishes where purposiveness does not slip in, though the writer may be unconscious of what he has done. Birds are said, even by the most mechanistic of ornithologists, to build nests in order to lay their eggs in them, and fish are said to swim up streams in order to spawn. The heart is compared to a pump that sends blood through the veins and arteries, and the lungs exist to provide oxygen as we inhale. To say this is no more heinous a methodological crime than to say that the sun rises and sets, though we know perfectly well that it stands still.

All three occupations obviously require at least a minimum of thinking. And thinking cannot go on unless certain prerequisites are present.

First, one has to have a clear idea of just what one's problem is. A problem arises when one observes a deviation from the rule. This formula has been in circulation for some years though it does not seem to have been universally accepted. If my definition, however, will be granted for the duration of this lecture, then we shall have to admit also that one must know the rule before one can spot the problem. This is not so easy as it sounds, for the gap between experience and the rule always exists, or, if the terms are not too distasteful, the gap between existence and essence cannot be eliminated. Consequently we have such concepts as tolerance in engineering or the acceptable error in science. Just how much of a gap we are willing to accept in all cases has never been calculated, for it varies in every type of investigation. In engineering the tolerance that is permitted depends on what materials are being used, what purpose a structure is to fulfill, what might be called the contribution of the measurement instruments, or better the organocentric predicament, the personal equation, the human equation, and no doubt a dozen other things. In laboratory science a technique has been elaborated to eliminate this gap to some extent, but never completely, for scientific laws when derived from experimentation are not even supposed to be absolutely true outside the laboratory. That is why the phrase, "Other things being equal," is always appended to them or understood. In other words they are always absolutely true on paper, just as statutes can usually be clearly understood on paper. The trouble arises when you try to apply them. With all this in mind, one can see that to state a problem clearly demands a lot more than simply to wonder.

In the second place, one has to know what evidence will be relevant to the problem in hand. There are innumerable things

that happen when anything happens, but no one would attempt to catalogue them all or even to suspect their relevance. In science, as in common sense, relevance is often established by tradition. But when a new sort of problem arises one sometimes has no tradition to follow. One of my colleagues at the Johns Hopkins University is interested in the incidence of congenital malformations of the heart. There was a time when the word "congenital" would have sufficed to stop any further investigation. If you were born with a defect, that was the end of it. It might indeed be the end in many cases. A child born today as a Mongoloid idiot can hardly become a mathematician, though it is possible that at some future date therapy will be devised even for that abnormality. But nowadays it is generally agreed that a child is nine months old at birth and that many things can happen to him while he is *in utero,* things that might be prevented. The nightmare of thalidomide made physicians realize that some malformations could be attributed to a drug taken in this case at about the twenty-eighth day of pregnancy. Phocomelia is not anything that can be cured, but it might be prevented. Now there are two customary attitudes to assume when confronted by something like this. One can follow tradition and announce that it is congenital, which it is, and thus incurable, or assume that as an abnormality it must be preventable. The second alternative is a metaphysical assumption, the assumption that nature is always regular and that the unnatural can always be obviated. But in the case of congenital malformations of the heart, one does not know a priori just what information about their uterine history is relevant and what is not, though the recorded histories of patients will reveal some. The scientist is in the position of Kepler when he framed his numerous hypotheses before hitting on the right one. I imagine that one would have to note down all the hereditary and environmental factors common to all children born with a cardiac malformation, even though at first sight some might seem to be beside the point. For, to take but one circumstance, even the economic condition of a family has influence upon the type of food a mother eats. And we know that deficiences in the diet of a mother can influence the anatomy of the fetus.

In the third place, one has to decide what kind of explanation will suffice, if one is seeking an explanation. Two kinds have held the field in the past (though there are others too), one called teleological, the other causative. The former are not very respectable nowadays, though in biology they are often slipped in when the scientist forgets his mechanistic principles. It would be next to impossible, I should imagine, to explain the dance of the bees as anything other than communication, and communication is not just mechanical. It makes more sense to say, as von Frisch did say, that the dance tells the workers of the hive where nectar is to be found rather than that it is a compulsory behavior pattern, like St. Vitus' dance.[1] There is always a third possibility, and that is simply to announce that whatever happens is due to the Will of God. This is irrefutable and, though of no practical value in remedying or changing a situation, it has sometimes been comforting. In the long run one has to stop and say, "These are the facts," and whether one identifies the facts with the Will of God or with what is called "the Case," the only difference is in the connotations of the two terms. As far as I have been able to discover, there is no rule for determining the limits of explanation. They seem by and large to have been determined by custom or tradition. A Darwinian of the old school is satisfied when he discovers that a variation is useful in the struggle for existence. Nowadays variations, in the form of mutations, demand further investigation.

These three items suffice to show that the kind of thinking that goes on is not the prerogative of either the scientist or

[1] Karl von Frisch, *The Dancing Bees* (New York: Harcourt, Brace and Co., 1955; also Harvest paperback).

the philosopher or the so-called man in the street. There is always a seepage from science into common sense, a seepage of both information and directives. One sees the former in the incorporation of technical terms into our daily vocabulary; the latter in our popular appraisals of behavior. People speak of the influence of heredity and environment, of analysis and synthesis, even of the fourth dimension, as if such terms meant the same thing in ordinary conversation as they do in science. Since the middle of the nineteenth century, terms like "dynamic," "creative," "progressive" are used as terms of praise regardless of what is powerful, created, or progressing. But at the same time the needs of society, such as general health and well-being, often determine what the sciences will take as their problems. I doubt that any biologist would deny the contributions of pharmacology to biochemistry or that any physicist would deny the contributions of aerodynamics to mechanics. But an analogous interplay can be found between science and philosophy. The Danz Lectures of Fred Hoyle are as much philosophic as scientific, and one of his main theses in that work, the thesis of cosmic unity, is, as he says, pure speculation. I have noticed with smug satisfaction that physicists like Eddington and Jeans, Einstein, and now Hoyle, when they get to the last chapters of their books cannot resist the temptation to extrapolate what they have learned into those regions that are as yet unknown. This is as it should be.

48 Philosophers of the present day are perhaps a bit too timid when confronted by such problems. Some maintain that they are only apparent problems, not real ones; others, less audacious, busy themselves with logical symbols, linguistic analyses, theories of meaning, and leave epistemology to the psychologists, cosmology to the astrophysicists, and ethics to the anthropologists. But they seem to forget that the methods used in these special fields are themselves problems for the philosopher, even when the conclusions reached by these methods are not. Scientists are proud, for instance, of what has been called the empirical or experimental method. But the concept of experience is in itself a very nebulous one, for it has meant a whole series of things, from pure and inarticulate sensory data—colors, sounds, and such—to the *consensus gentium,* and has sometimes included our feelings and intuitions. When a scientist insists that truth is the character only of interpersonal experiences, he leaves to the philosopher the problem of what should be excluded from any individual's experience in order to make it interpersonal. In one reasonable sense of the word "experience," everything that a man undergoes twenty-four hours of the day—his emotions, fears, aspirations, likes and dislikes, his dreams and illusions, his

repulsions, his conversations, his reading—makes up his experience. Much of this, as I intimated in my first lecture, has to be eliminated when one is talking science. But it is eliminated in order to make science possible, not because it is nonsensical or unrelated to knowledge. There has always been an aesthetic element in science which is not revealed in the content of any experiment but exercises its influence well before an experiment is undertaken.

Take, for instance, the words "the Cosmos" or "the Universe." I think I am right in saying that there is attached to these words the notion that everything that exists forms some sort of unity. But unity is a very vague word. It can be either the unity of stuff, the stuff of which a collection is made, or it can be some formula that states the composition or structure of the collection. It sometimes means a unity of origin and sometimes a unity of end or purpose. I gather that in Professor Hoyle's theory it is a material unity essentially, and that, whatever this material is, it is the origin of everything that exists. If this is true, and I can only conjecture that it is, then the philosopher will want to know why things seem so different from one another. The difference among the various chemical elements is not only in their atomic weights, but in their behavior. Common sense, like chemistry, agrees that you cannot substitute one element for another, that you cannot use helium for oxygen or sulfur for silver. Hence the philosopher would insist, I hope, that material similarities do 50 not satisfactorily explain these differences in behavior. Is one to announce a universal law of material differentiation, as one might say that it is a differentia of life to split up into genera, species, and varieties? A speculative philosopher could have a very amusing time some rainy afternoon with this sort of game.

He might even go to the length of suggesting that the use of the concept of a universe is taken over from art, where aesthetic unity has always been appraised more highly than diversity. One does not construct a unit simply by saying that every particle of matter that exists influences in some way every other particle of matter. For what precisely is this influence? For Newton it was gravitation. But Newton in the General Scholium went a bit further and said that, since the Law of Gravitation was *one* and applied to all matter, it must express the will of one lawgiver. This metaphysical invention did not, however, account for other singularities which are familiar to all of us; it accounted only for the way things move. To think of things as essentially masses in motion is to forget accidental properties that are even more noticeable than their masses and velocities. I refer of course to their perceptual properties and their goodness and badness, their beauty and ugliness. These qualities may be correlated with the velocities of the elementary particles of things, including among the things animate bodies. But why out of their collisions these nonmechanical properties should arise, is still obscure. To think of things as essentially masses in motion would seem to be to utilize the metaphor of the artisan who takes bits of matter and moves them about in space to construct something out of them. And this in turn is to transfer to a field in which life has no place something that is peculiar to life. One might just as well revert to those ancient cosmogonies in which everything was produced from a cosmic egg, or to those in which a primordial couple gave birth to everything. For when you enter this area of thinking you are forced to introduce some figure of speech. A man like Aristotle was able to conclude that the world had no beginning or end. He may have been wrong, but at least he was not taken in by Greek mythology.

Whatever the scientist may do when confronted with a statement that contradicts habitual ways of thinking, the philosopher has to consider it seriously. He has no super-philosopher to whom he can say, "This is your problem." His professional ethics makes him ask whether it is necessary that all things have a beginning and an end. The only things we know anything about that do have beginnings and ends are the things that we make and destroy and also living beings. But in both cases, as some of the ancient Greek philosophers

saw, it is only their composition that begins and ends, whereas the stuff out of which they are composed existed prior to their origin and continues to exist after their destruction. If, then, we were to be strictly empirical about the universe as a whole, we should not ask the question of how it began but rather of when it was put together, or when its present organization began. We should probably ask the wrong kind of question here, too. I have no competence to discuss the implications of such postulates but can only point out that if one says that each star is part of the whole, then a very precise definition of "part" and "whole" should be demanded, for both terms, as I have suggested, are ambiguous. Moreover, since all of these meanings are derived from experienced things, when they are applied on a scale much larger than any possible human experience, they become metaphorical. And nothing is more fallible than arguing from metaphors.

Science also influences philosophy through its humanistic implications. It has sometimes been said that if there is universal determinism, so that everything that happens has a discoverable cause, then what men do is caused and they have no power over it. In short, people who say this also say that ethical responsibility is an illusion. This clearly does not follow unless causes are all inoperative, in which case determinism is nonsensical. For if determinism is really universal, then human beings must also determine some effects, and the mere fact that they are also subject to causal influences is irrelevant. Causal laws, like all laws, describe classes of events and apply to the individual only so far as he is a perfect specimen of the class to which he has been assigned. The assignment, be it noted, is made by man, not by what we are accustomed to call Nature. Within every collection of individuals there is deviation. And we know that no matter how strong an incentive may prove to be on the whole, there are always some individuals who are unmoved by it or who resist it. There is no point in being deluded by the common nouns that we use. No man is just a specimen of *Homo sapiens* or of any of the other classes to which he has been allocated. He is always himself. This is no truer of men than it is of anything else, cabbages or goats. For outside of laboratories, let us repeat, there are no pure specimens of any class. And inside laboratories they are purified by fiat. Moreover, when things are acting in a complex, they often behave differently from the way they behave in solitude, assuming that you can get anything into solitude. Even the chemical elements behave one way in isolation and another when in composition. Sometimes the very same elements, as in the carbon compounds, form substances which differ only structurally, and that difference lies in the spatial arrangements of the atoms in their molecules. Yet

of these structures one cannot be substituted for another. Though ribose and deoxyribose are almost alike, the latter having only one less oxygen atom than the former—surely nothing to make a fuss about—the absence of that one atom means that deoxyribose does not have the same effects as ribose. Any elementary textbook in organic chemistry will illustrate the point that structure is as determinative of what a thing will do as substance is. You thus can have the same substances in two different things and yet find that each one will behave in a peculiar way because of the organization of those substances.

Consequently, the fact that living beings are composed of hydrogen, oxygen, phosphorous, potassium, iodine, and so on does not permit us to conclude that they will behave as those elements do, even though you can burn up the human being and recover the elements of which he is composed. If he is deprived of oxygen, he will die; there is no denying that. And dead men do a lot less than telling no tales. But if we grant that human beings can act as units, not as collections, though sometimes they do not, then you cannot infer their behavior from the behavior of their parts or constituents. Nor on the other hand can you deny that there are general descriptions of the way they will behave when they form units in large collections. But such general descriptions have to take into account the way every individual in the group behaves. Otherwise they are simply not true.

This is to be sure a commonplace, but one that is neglected. It is the commonplace that every individual contributes something to the events in which he participates. If you hesitate to call this responsibility, then some other name that means the same thing will have to be found for it. If we mean by responsibility the power to do anything whatsoever, regardless of our anatomy, our state of health, our position in the family and in society, our wealth or poverty, our knowledge or ignorance, and all the other complexes into which we fit, then I should imagine that we are talking nonsense. There are limits to responsibility as there are to all forms of behavior, and we have to stay within these limits. A falling body will not accelerate as the books say it should if it bounces off an awning or drops into a net. A bottle of arsenic on a shelf will poison no one. Such sentences are obvious. But their implications do not seem to be obvious. Causation, I am saying, occurs in contexts, and everything in a given context may influence the outcome. If human beings are part of the context, they too must have their share in the outcome. The job of the scientist is to discover just how responsible human beings can be, and the job of the philosopher is to tell him to. He cannot evade this job by talking about

universal determinism.

Sometimes scientists when they become philosophical, and philosophers when they think they are scientific, talk about the various causal series that affect human life as if each in itself and in isolation from all others determined human history. For instance, biochemistry, economics, the evolutionary process, and even astronomy have been called upon to reduce humanity to a level of impotence. This is the impotence that comes from fear. Unfortunately for such people, we are not merely biochemical specimens, economic specimens, maggots on a dying planet, but a combination of all such things and others too. There is no saying that biochemical structures such as ours which have a history, which try to satisfy their economic desires, which live in communities, and all the rest, will act as if they were only one of these various things. One might just as well argue that a man can breathe under water since water is one part oxygen, or that a match will burn in water since water is two parts hydrogen. Analogously one can argue that, since men have to eat in order to live, getting food is the determining factor in human behavior, or that, since man has to get along with his fellows if he is going to survive, cooperation is the determining factor.

It is clear that such inferences are not true, for they run counter to observation. Nor does recourse to the unconscious in such matters help, for if an unconscious motive can be effective only by disguising itself, then it is the disguise that is more effective than what is behind it. Finally, there is no need to resuscitate the will as a sort of engine that propels man on his various adventures. I am talking about the whole person, in so far as he is a whole. A man does not walk abroad carrying concealed within him another man who does all the work.

None of this is a plea that the philosopher should proceed without respect for the sciences. For no matter how lofty our intentions, we are forced to consider those limiting conditions, as defined by the sciences, which alone will keep us sane. When philosophers attempted to construct an anti-materialistic metaphysics, they were frustrated by the stubborn interference of physical law. From Bishop Berkeley down to the cruder pragmatists, they have had to make room for a world that is not to be described in terms of psychical processes alone. If David Hume had been right and the world just a theater of passing impressions and ideas, the question of an external world would never have arisen. If there were no self, for instance, why should we raise the question of its existence? If we had no reason to believe in the existence of other minds, why should we think we ought to explain the illusion of their existence? In brief, if we accept the conclusions of such philosophers, then a number of philosophic problems that have pestered the human mind since Locke could not have arisen. We have outgrown, I suspect, the feeling that we ought to explain in chemical language why water in a dream does not put out fires, or in terms of biology why people in a dream can turn into other people right before our eyes. We leave such matters in the hands of psychologists, admitting the discontinuity of the dream world and the waking world. So the man whom I have called the crude pragmatist has to pose the problem of why some ideas that we frame do not work well while others do, and when he does ask this question, he cannot resort to anything other than a nonhuman world that thwarts his desires. The philosopher will always find out as much as his education permits about the world as interpreted by the sciences, and he will draw his conclusions from what he has learned. But in so far as he is interested in synthesizing the information for nonscientific purposes, he cannot rely on one science alone.

The nonscientific purposes of which I speak are religious, ethical, aesthetic, political, and any other that concern the values which men would like to secure. In the old days these were called the true, the good, and the beautiful. Whereas it is true that the beauty, the elegance, the completeness of a scientific theory has value in itself, science is not supposed to be produced for the sake of such value. For sooner or later the scientist will want to apply his conclusions if only to see if they can be empirically verified. But the philosopher wants more than that. He wants to know, for instance, whether it is worth while trying to make a better life for his fellow men if all life is to be extinguished by nuclear bombs or poison gases. Since I am in the privileged position of being asked to expound my own views, I should say that it makes no difference what the future holds in store for us when we are trying to ameliorate the present. If what we are seeking is something good, let us have it if only for one generation. The uncertainty of survival is just as great when walking along the streets of a modern city as it is in a jet-propelled plane. Nor is a good any the worse because its duration is not very long. The foreseeable future is short in any case, and the probability is that men will always use good means for evil ends. But I fail to see why, for instance, the beauty of this site surrounding us becomes ugly because at some future date it may be burned to a crisp. Similarly I cannot understand why our hopes are silly because we are nothing but a conglomeration of cells whose individual characters are determined by the nucleic acids. The important thing is that these conglomera-

tions can have hopes and, what is more, can work to realize them. Is one to argue that the starry sky is ugly because the
60 dots of light are only incandescent gases? Incandescent gases just happen to be beautiful. One might just as well say that a painting is not beautiful because after all it is nothing but linseed oil and some coloring matter.

The values of which we are in search are not to be measured in terms of their source, of their material roots, or of their duration. The use of science to dispute this is futile. Our hopes, to be sure, are often foolish; this must be granted. But that we have hopes is a fact. And that sometimes these hopes can be fulfilled is also a fact. In the third place we can learn how to distinguish between the foolish and the wise ones, and in the fourth place we can find out how to fulfill the latter and discard the former. But I doubt that simply lying back and letting Nature take over, as if we were not an integral part of Nature, will ever do more than reduce us to the position of those animals, like the reptiles, who seem to have given up the struggle for existence and to have hidden in dark holes or covered themselves with a shell. In my youth we were all told that something called Evolution with an initial capital was going to solve all our problems. Everything was going to get better and better whether we did anything about it or not. But we soon woke up to the fact that there was evidence of deterioration in evolution as there was of amelioration, that there seemed to be a swing between submission and aggression, though these terms are fanciful. In any event we now know enough of human history to have learned that we cannot count on natural forces alone to achieve the good life, however we define it. Of course if whatever is, is right, then I am wide of the mark. But the ideas of better and worse are compelling ideas, and we can no longer consider Nature as a system in which humanity is missing. I should guess that any scientist would grant that, as far as our desires are concerned, they can be fulfilled only by us and not by the landscape. There have been cultures in which the sources of energy have been present and they were not utilized. The presence of fuel oil in the Middle East was not sufficient to create industry, nor was the absence of land in the Netherlands enough to prevent agriculture. In fact the history of civilization is also the history of man's struggle against Nature, not merely of his submission to it. If European man had submitted, his descendants would still be inhabiting the caves along the banks of the Vézère. In the first series of these lectures, Julian Huxley spoke of the need of limiting the population, a need which most educated people would not deny. Though he is a biologist, he did not say that we ought to let Nature take its course and that the surplus population could be allowed to come into being and then be decimated by disease, starvation, or perhaps cannibalism. That would

have been the course plotted by those who preach man's incompetence over the course of history. It would be possible to reduce the population by dropping a few low-yield atomic bombs, if there are such things, over those areas which are overpopulated. I do not imagine that the land would be good for much for some years later. But about that I can give no reasonable opinion.

The phrase "the impact of science" is found in the bequest on which these lectures are founded. And surely one result of this impact has been massive terror. Philosophers, instead of spending their time on manipulating pseudomathematical symbols, would have done better to tackle the problem of fear, as Lucretius did, and to see as clearly as possible how justified it might be. It is true that the scientists themselves are not responsible for fear. The responsibility lies in the hands of the governments, journalists, and amateur preachers. When a tool is invented, its use will be determined not by the inventor but by the rest of us. Alexander Bell cannot be blamed because telephones are used to interrupt meals, to gossip, to obviate the need of putting promises down in writing. Nor can Gutenberg be blamed because printing is used for lies, slander, and other forms of chicanery. There is no natural law, as far as I know, that compels us to misuse our blessings. We even misuse our political rights, such as freedom of speech. But the fact remains that we recognize such abuses and that some of us have the courage and the decency to fight them.

Along with the widespread fear of which I have been, shall I say, whimpering, there is widespread inertia. And that, too, is caused by the impact of science. Such doctrines as economic determinism, for instance, which by the way was not upheld even by Engels,[2] reduce the young to impotence. What is the good, they say, of studying hard, of working for a better life, if it all amounts to nothing more than stuffing the wallets of those whose wallets are already bulging? Similarly, if all our aspirations are masked sexual appetites, then one gets the idea that patriotism, brotherly love, learning, the arts, and religion are hardly worth the effort. One can gratify one's sexuality much more easily than by working so hard and sacrificing so much. It would be just as easy and more valid to conclude that if economic interests and sex can give rise to such values, then the more power to them. The fallacy is the old one of identifying the origin of something with the thing itself, and of denying to the end the values that are

[2] See his letter to J. Bloch, of September 21, 1890. Translated in Sidney Hook, *Toward the Understanding of Karl Marx* (New York: John Day Co., 1933), especially p. 333.

absent from its source. But this is about as reasonable as saying that a chicken cannot have feathers because an egg has none. I doubt that any philosopher would draw such fallacious inferences, but they are common enough in magazine articles to be a powerful depressant on those who read them. It is, to be sure, also true that men and women are discouraged not merely by pseudophilosophical arguments, for there are plenty of other influences tending toward the same end, influences such as international slaughter and internal dissension. The spectacle of racial conflicts in parts of this country, for example, is nothing to make one rejoice. But I might in fairness add that it is the young who have set an example of moral courage to their seniors, not merely in helping in the battle for civil rights but also in international aid.

We have by now almost reached the frontiers of philosophy. In the long run the question of faith arises, for the choice of premises is always a matter of faith. And in philosophical matters the problem is close to religion. We shall in our final lecture discuss the challenge which the sciences make to that.

The Challenge to Religion

Discussions of religion may center about three themes—religion as a personal experience, as a set of theological tenets, and as a society of communicants. The importance given to these three separate themes varies from person to person. Nevertheless the connection among them should not be neglected, especially if one is trying to be just, for the religious experience is interpreted both by theologians and by the clergy; the theology itself is presumably based on the experience of someone who has had an illumination or revelation, if only twenty centuries ago; and the church itself is a society in which both the experience and the theology are vital principles. There are men to whom theology and ecclesiastical organization are unimportant; but that is often, if not always, because they have not pushed their religious beliefs as far as a philosopher would like them to go. There are undoubtedly others who think that the cult is of the greatest importance and that the revelation to individuals may come through constant and loyal observance of the cult. That was Pascal's thesis.[1] A child is usually indoctrinated into religious observances long before he can understand the meaning of their theological bases, but that has never prevented some adults who have gone through this indoctrination from being fervent believers. One can also be a thorough student of theology and yet never have had a religious experience. It is not essential that one decide which of these three elements in religion is the most important. One can maintain the thesis that if one is lacking, then a man's religious life is incomplete.

Historically it is probable that ritual preceded its rationalization, or, if the terms are not offensive, that cult preceded myth. We learn our American history from practice, not books, from celebrations of the Fourth of July, Washington's

[1] "Vous voulez aller à la foi, et vous n'en savez pas le chemin; vous voulez vous guérir de l'infidélité, et vous en demandez les remèdes: apprenez de ceux qui ont été liés comme vous, et qui savent ce chemin que vous voudriez suivre, et guéris d'un mal dont vous voulez guérir. Suivez la manière par où ils ont commencé: c'est en faisant tout comme s'ils croyaient, en prenant de l'eau bénite, en faisant dire des messes, etc. Naturellement même cela vous fera croire et vous abêtra (*Les Pensées*, ed. Pléiade [Paris, n.d.], No. 451, p. 1215 f.). Cf. No. 450, p. 1212.

Birthday, and Thanksgiving, by singing the "Star-Spangled Banner," by pledging allegiance to the flag, long before we read books explaining these observances. And similarly we say our prayers before we know anything about the nature of God. Just as the Twelve Apostles came to Christianity after a lifetime of Judaism, so the adult Christian comes to an understanding of his belief after a lifetime of Christian practices. There is nothing peculiar about this. We live before we are able to reason, and a good bit of reasoning is spent on interpreting what we do automatically. If we did not do these things automatically, we should feel no compulsion to give reasons for them.

Moreover, if our beliefs are to have any efficacy over conduct, they must be taught after the conduct in question has become ingrained into a person, not before. You can argue with formal doctrines; you cannot argue with acts. Acts give rise to beliefs, and some beliefs have no effect upon acts. Even when the beliefs in question spring from commands, such as statute laws and ethical commandments, their bases may be explored to a point at which only fiat can justify them. I have yet to find a demonstration of the basic principles of ethics which cannot and has not been disputed. On the other hand, I have yet to find a list of virtues and vices which would not be accepted by almost everyone, regardless of whether he strives to exemplify the virtues and to avoid the vices in his daily life.

If I am right about this, then science can have little influence on religion as practiced, though it may have some influence on theology and ecclesiastical organization. One has merely to think of the problem of extirpating bad habits to see how little effect talking has upon them. Habituation to alcohol and tobacco, even to gambling, is supposed to be bad, and no end of sermonizing has been spent trying to eliminate them. But so far it has all been wasted effort. Some of us can remember the "noble experiment" of Prohibition, the result of which was to prove that people will violate the Constitution and a federal statute to do what they are accustomed to doing. As far as religion is concerned, for the last two thousand years men and women have recited the Creed in church, gone home, and then forgotten all about its moral implications. The pageant of wickedness which is human history was not staged exclusively by skeptics. The ethics of the Sermon on the Mount admittedly include those precepts of which the world is most in want; moreover the people who regard them as such believe that they were the words of incarnate God. Surely I need not say much about how far these precepts have guided Western society.

Yet even if our theology follows from our religion, it is a necessity for a rational man to want a theology and to satisfy that want. As I have just said, if you pick up a book on ethics, dealing with the concepts of right and wrong, good and evil, motive and consequence, egoism and altruism, inherent and instrumental values, all you get is new reasons for doing what you already believe to be right. There are, of course, a few writers like Nietzsche who have turned things upside down and made virtue into vice and vice into virtue. His ethics as a matter of fact was a pretty good description of what men do, but it was shocking as a proposal of what they ought to do. At least this is what his critics said. The devaluation of all values was perhaps too accurate a picture of human behavior to appeal very strongly to the majority of ethicists. They were more interested in the *ought* than in the *is,* and what ought to be seldom is. At least it seldom is now, though it is hoped that it will be in the future. May I express at this point my wonder that human beings can have so much confidence in time? Why do they think that our natures will change simply because a new generation shall have come into being? That generation will also be human. If individuals as they grow older do not become any better than they were as children, in spite of experience, why should their offspring be any better? The hope is similar to that expressed in literary criticism when the critics tell us that we must wait for the judgment of posterity before we can say whether a book is good or bad. In the first place we shall be dead and will not be able to hear what posterity has to say, and in the second place 70 posterity will be simply some more men and women.

This is not to commit us to the belief that human nature never changes. I am not sure just what part of us is our nature, but if it is that part preoccupied with standards of behavior, then of course it has changed and it may change again. It changed fundamentally, as far as ideas were concerned, with the acceptance of Christianity, and it changed again after the French Revolution. Men's attitudes toward their fellow men changed radically when they could admit that they had social duties. Their attitude toward social rank and privilege changed in the nineteenth century, and one has a feeling that it is changing now as more and more people see the effects of war. As the historian knows, the seeds of these changes appeared well before the changes themselves became common; even the Church Fathers granted that ancient philosophy was in part a preparation for the Gospels.[2] But I must not digress.

[2] As a matter of fact, Father F. C. Copleston in his *History of Philosophy,* Vol. I (New York: Doubleday and Co., 1947), assumes the same principle. It is historically true that Stoicism and parts of Platonism, especially the *Timaeus,* did influence Christianity in the sense of making it easier for pagans to accept certain of its tenets: e.g., the brotherhood of man and the idea of divine law. But that is not the thesis of the Fathers. They attribute the role of ancient philosophy in preparing the way for the reception of Christianity to God's plans for what Lessing was to call the education of the human

I have said that what we demand is rational grounds for believing what we believe. Belief, then, can be prior to reason. This is summed up in St. Augustine's phrase, *Credo ut intelligam*. First, for example, you believe in God, and then you try to understand why. In other words a man is confronted with a situation, and this he cannot explain away as a dream, a hallucination, or an illusion. He therefore sets out to see how far he can justify its existence rationally. If this seems farfetched, think of the reverse situation in which you are faced with something in whose reality you do not believe, let us say telepathic communication or ghosts. You immediately use your reasoning powers to prove that your disbelief is right. There just cannot be telepathy or ghosts, you say, and all the reports of telepathic messages or ghosts are discounted as impossible prevarications or deceptions. The immediate experience always has temporal, if not logical, priority over its interpretation, and whereas the interpretation may be wrong, the sting of the experience carries conviction, even though it is in essence inarticulate. When the mystic reports on his beatific vision, the rationalist will argue himself hoarse trying to refute him. But the mystic replies that his experience was his own private experience and that no one else can tell him what it was like. And when it is replied that

---

race. Ancient philosophy thus played a part analogous to that of the Old Testament, but of course far from identical with it.

you can get the same result from eating peyote, the ready answer is that eating peyote is one of the ways of having a beatific vision. When Jean Cocteau was uneasy about having been converted while under the influence of opium, Jacques Maritain replied that maybe God had used opium to bring about a conversion that was not likely to occur otherwise. When you have a direct experience of any kind, you know better than anybody else what you are experiencing. Nothing could be less plausible than that colors could be caused by the interaction of light rays and the optic nerve, neither of which has the color which one sees. The question thus is whether your interpretation of your direct experience is correct, which can only mean "acceptable to science." People have been known to be wrong even in their sensory perceptions, though no one could reasonably deny that they had them. The desolation of the child who first discovers that his father is not infallible, of the mother whose son has lied to her, of the wife who thought she could trust her husband—these are honored themes of novelists and dramatists, and all of them illustrate weakness of belief when not fortified by evidence. But they do not prove that evidence suffices to cause belief. Before it can be utilized, one has to suspect that it is evidence *of* something. I once gave a lecture to an audience in which there were two people in the middle row who seemed to be much more interested in each other than in the lecture. It later turned out that one of them was stone deaf and the other was translating my words into the hand sign alphabet.

In mathematics, which of all our thinking is most detached from experience, we nevertheless start out by saying, "To prove such and such. . . ." We do not immerse ourselves in a set of theorems and let them do our work. Somehow or other the idea strikes us that if P and Q are true, then R ought to be true also. And we set to work to see if it is true or not. Why this expectation strikes John Doe and not Richard Roe, though both may have the same education as far as two different people can have the same education, can be answered only by the psychologist. There is a large literature on the chronology of discoveries and inventions, and all that has been found out is that several of them have been made almost simultaneously.[3] But the famous cases of Leibniz and Newton, Darwin and Wallace, simply suggest that something must have been in the air to turn their thoughts to the same problems and suggest the same answers. What is wanted is more

[3] See, among other articles, R. K. Merton, "Singletons and Multiples in Scientific Discovery: A Chapter in the Sociology of Science," *Proceedings of the American Philosophical Society*, CV (1961), 470; W. F. Ogburn and Dorothy Thomas, "Are Inventions Inevitable?" *Political Science Quarterly*, XXXVII (1922), 83.

knowledge of Newton, as contrasted with his fellows at Trinity College, and analogous information concerning the others. But that is no longer available, if it ever was. There is an eloquent description by Philo Judaeus of the experience of getting a new idea. This experience, he says, was so wonderful that it must have come from God. The description runs as follows:

> There have been times when I have wished to write according to the custom of philosophers, and have known accurately the essentials of what I was to put together, yet I have found my understanding impotent and sterile and have given up my work as useless, cursing my self-conceit and awe-struck at the might of the Being by Whom the womb of the soul is opened and closed. But there have been times when I have come to work empty and suddenly become full of thoughts falling like snow from above and invisibly sown, so that under the influence of the God within me I have been seized with corybantic inspiration and have lost consciousness of everything, the place where I was, those present, myself, speech and writing. For I received messages, conceptions, the rewards of illumination, and a very clear perception of things such as would arise from the most vivid presentation of things through one's eyes.[4]

No one has the right to doubt Philo's description of what happened on such occasions, although one can reasonably

[4] *De migratione Abrahami*, VII, 34. For a closer translation, see that of F. H. Colson and G. H. Whitaker, in the Loeb Classical Library, Vol. IV (Cambridge, Mass.: Harvard University Press, 1932), p. 151. An interesting parallel, but without the supernatural implications, will be found in William Barrett's contribution to a symposium on *Religion and the Intellectuals*, published by the *Partisan Review* in 1950 and entitled *PR Series* number 3. See especially p. 38.

doubt that it was induced by God. But how would one find out?

Clearly the only investigators capable of answering such a question would be psychologists. There are several experiments that they might try. They might try to induce the same condition in a subject by means of drugs or by hypnosis, controlling the experiment by the usual methods, and they might conclude that since the same type of experience could be produced by them under laboratory conditions, there would be no need to assume the intervention of God. This, however, would convince no one who was not already skeptical. For he could always reply as Maritain replied to Cocteau. Philo Judaeus was not proving the existence of God, in his case the God of the Old Testament, for he already believed in His existence. He was describing a case of divine help in the acquisition of knowledge. Now it is hard for a lecturer who has always had to work to gather his materials and then to organize them, to report on inspiration, but he can at least report on the sudden flash of what he thinks is insight, on what he believes to be the telling example, on the clearest way to state a conclusion. How this type of thing comes about he does not know, though in his case he does know that it is not a revelation. But if he were convinced both of the transcendent importance of his own ideas and of the complete absence of preparatory struggling to clarify them, he might after all attribute their presence to divine

inspiration. Newspaper and radio commentators often write as if they had a private wire to the Holy Spirit; they reveal secrets, make prophecies, denounce abuses of power, much as Jeremiah might have done in their place. Who knows what they believe about the source of their gnosis? And if, as I suspect, they think it is supernatural, would they be convinced by any psychologist who tried to demonstrate its earthly origin?

In so far as religion is an attitude based on sudden flashes of insight, I do not see how science could modify it. It is too private a matter to be adequately verbalized. But the religious attitude is more than that, too. The astronomer who stares into interstellar space is in the same posture as those men whom Seneca describes as our primordial ancestors gazing awestruck at the nocturnal sky. But I doubt that the awe is peculiar to savages; the sight is enough to shake anyone. Similarly the biologist who thinks of the whole chain of life from unicellular animals to man, or who meditates on the evolution of such organs as the human eye or brain, can hardly avoid an analogous feeling. The spectacle is bound to seem more like a plan than a random hit-or-miss collection of incidents. Yet one also knows that plans are human events, and one is always on the brink of reading humanity into the natural order. The biologist should, I suppose, avoid such opinions and should maintain stubbornly that a flatworm in the natural scale of creatures is as good as a man. No one wants to be a flatworm, to be sure, but apparently flatworms do not mind. If, however, one goes over the brink and believes that there is something inherently better in being a man, then one has to read into the evolutionary process a movement toward the better. There is no way of avoiding this, granting the premise. I admit that this is not what is normally called a scientific problem, but it is a human problem, and one cannot be a scientist twenty-four hours a day. Once in a while, if only in sleep, human interests slip in.

When they do slip in, another problem arises which, as far as I know, no scientist has ever treated seriously. And that is, What is the status of intellectual, aesthetic, and religious preoccupations if man is only a biochemical, zoological, and economic item? These interests may be aroused, for all I know, by the functioning of our endocrine glands, our metabolism, our digestion, but why should our glands, or whatever takes their place here, be busied turning out symphonies, poems, hymns, and philosophic treatises? These things are of no use to anybody, except possibly to the owners of wood pulp. No one ever got on any the better from either producing or enjoying them. Yet almost all men indulge in them, not merely the intellectuals. Every county fair used

to show patchwork quilts, paintings, burnt-wood picture frames, jigsaw knickknacks, pictures made out of postage stamps, all fabricated by the rural population. These people had no aesthetic programs and were not trying to sell their products. They just thought the objects were beautiful. There may well be a wide aesthetic gap between such things and the works of professionals, but to the people who made them they were admirable and a justification for spending time on them. If you ask what biochemical or biological value resides in such work, you will not find the answer in technical scientific treatises. And if the reply is that there is no such value in them, then you have to conclude that alongside of that which is explicable in terms of natural science, there exists something which is nonscientific, though not necessarily antiscientific.

Leaving out of the discussion the source of religious insight, it may be well to point out that religion itself, as a tradition, has been the greatest source of aesthetic value that exists, both in the East and in the West. Organized religion has certainly done a great deal of harm. The Church has been cruel, intolerant, an obstacle to intellectual invention. Nor has the life of ecclesiastics always been edifying. Nevertheless the ideas on which the Church has been founded have inspired architects, sculptors, painters, composers, and poets to produce works of art incomparably superior to those inspired by science. Newton may have demanded the Muse, but try now to read what that Muse dictated. Try to read Erasmus Darwin on the loves of the plants, John Armstrong on the art of preserving health, or John Dyer on the raising of sheep.[5] The drawings of Vesalius have often been highly praised, but compare them with the nudes on the ceiling of the Sistine Chapel. Not even patriotism has produced paintings equal to those of Giotto, Piero della Francesca, the brothers Van Eyck, and almost countless others. And so far as literature is concerned, only love has been a theme more stirring than adoration. To continue in this vein would be to tumble into a slough of sentimental rhetoric, and I shall spare you that. I shall be satisfied to say that similar remarks could be made with equal force about the cultures of ancient Greece, India, China, as well as of Southeast Asia. A civilization without art, moreover, would be intolerable. And it is next to impossible to believe that the men who made Angkor Wat, or composed the Greek tragedies, or wrote such masses as Bach's in B-minor or Mozart's *Requiem,* were simply playing, fiddling about with

[5] These men and others will be found quoted in *The Stuffed Owl, an Anthology of Bad Verse*, selected and arranged by D. B. Wyndham Lewis and Charles Lee (enl. ed.; London: J. M. Dent, 1948).

words and notes and bricks to occupy their leisure.

If such things come out of religious feeling, then our problem is not so much that of the impact of science on religion as of religion's impact on science. As a matter of fact, the sacred texts, at least in the Occident, have posed problems to the scientists in the past, as in the case of the origin of species or of geological dynamics. The exegetes have clearly done their best to frustrate scientific discovery. Yet by insisting on the scientific significance of their texts, they have forced scientists to examine their own theses more closely. It may be true that the belief in creation *ex nihilo* could not stand up under geological criticism; there are dozens of texts which even the most pious have had to interpret allegorically, so contrary are they even to common sense. But oddly enough Professor Hoyle has to resort to creation to solve one of the problems of the expanding universe. When asked, Where does "the created matter come from?" Hoyle answers, "It does not come from anywhere. Material simply appears—it is created. At one time the various atoms composing the material do not exist, and at a later time they do."[6] There is no mention here of a Creator, and there may be some question of the appropriateness of using a term that comes from

[6] I quote from the paperback edition of *The Nature of the Universe* (Mentor Books, 1955), p. 112.

theology in a nontheological context. But, nevertheless, resorting to such a term illustrates how we are all bound by tradition and how limited we are when we come to basic problems. For something to appear *ex nihilo* is either inexplicable or it is simply a metaphor. I doubt whether the problem would have arisen at all if the astronomer were not bothered by the notion that anything that appears must have come from something or somewhere. In brief, one extrapolates by analogy an empirical situation into the unknown. But maybe that is what the author of Genesis was doing too. Who knows?

I have no way of knowing whether Professor Hoyle feels the religious connotation of what he has written. Nor does it make much difference whether he does or not. For him the problem is predominantly a mathematical one. But to the outsider the idea that matter is constantly being created so that the universe does not disintegrate is one profoundly evocative of religious feelings. He cannot resist the impression that this is simply Genesis in a new language. He may be wrong about this. But that is irrelevant. Most of the statements of cosmological fact in the Bible would be admitted to be literally wrong. But when men want to believe something, they can always turn statements of fact into symbolism. The Greeks did it with their mythology, and the Christian Fathers, preceded by Philo Judaeus, did it with the Bible. What interests us here is not what they believed so much as their

desire to believe. It is not merely cosmology that is involved. The behavior of human beings is just as mysterious as the origin of the galaxies. To make sense of what we do is next to impossible, if you mean by sense a law that will describe precisely what any human being will do on every sort of occasion. We have the suspicion that all events can be made intelligible, as if there were no such thing as chance. One of the most presumptuous opinions that men have uttered is that chance is simply human ignorance. But even if that were so, there are two kinds of ignorance, that of things which we do not happen to understand as yet, and that of things that are impossible to understand. Now chance events are of the latter kind almost by definition. You can calculate probabilities, but every freshman knows that such calculations do not apply to real events except *on the whole*. To deny this is to fall into the Gambler's Fallacy, namely, that if the probability of something's happening is one out of ten, then for the first nine times it ought not to happen and on the tenth it must happen. In such cases only observation will tell the truth, and prophecies fail. On the basis of probabilities alone, it must have been next to impossible for any of us in this room to be born, let alone to gather here today. There is no way of knowing just which spermatozoon will penetrate a given ovum in normal conceptions; that is a matter of chance. It is perhaps true that some day fathers will be obsolete—which would blow the wind out of the Oedipus complex—and spermatozoa manufactured in laboratories. Similarly, I suppose, the same could be done for ova. But that would not eliminate the fortuitous element in conception as it now proceeds.

We have insisted in these lectures that the events which go on in time and space are different from those in the books. The difference involved is a logical surd. That surd is called existence. Some of the Church Fathers used to raise the question of why God should have created the world, since, after all, things were all right in Heaven and He needed nothing to perfect His nature. The modern philosopher raises the question of why anything whatsoever should exist, since the conceptual world of the scientist is more orderly, more beautiful, and a lot less trouble. The scientific world reflects a selected group of the existent world's traits and organizes them in consistent theorems. This, the scientist thinks, renders the existent world intelligible. But just as a photograph of a person shows only one side of him in a contrived illumination, so a given science cannot be asked to do more than show a set of patterns derived from and not identical with the existent things in which it finds them. In both cases the pictures may be much more beautiful than the models. It is also to be noted that no praise or blame attaches to the models if they do not

live up to what the pictures demand of them. The blame is put on the photographer or the scientist. The former is blamed for flattering his subject, the latter for errors in calculation or laboratory technique. We tend to think of the natural world as just lying out there, waiting to have its picture taken.

But human beings do not live in that beautiful world. They admire it when they see it, and of course they forget that life is not a scientific laboratory. The regularities of the world of science are abstracted out of the world we live in; that is undeniable. But we do not live in a world of regularities. The unexpected is so usual that when things come out as we anticipate, we are surprised. That is why we turn to interpretations of experience that are nonscientific. The degree of irregularity seems to vary from the orderliness of the solar system, with its recurrence of day and night, the seasons, the rising and setting of the various planets, to the vagaries of human nature. To relate such irregularity to something like human caprice does not appear to be unreasonable. Heine was not alone in calling God an Arch-humorist. If one remains on a purely human level, forgetting all one knows about the natural sciences, irony would indeed appear to be the secret of history. The only sense one could make out of it would be that we are the victims of a tremendous practical joke. Think of the occurrences of history as rewards and punishments. One cannot undo what has been done by punishing a malefactor. And as for reforming one by punishment, we may merely arouse more intense resentment in him and the desire for revenge. As for rewards, it must be pleasant to get them, but the only pure reward is self-approbation. I have often read essays saying that the pagans and the ancient Hebrews thought of their gods as essentially jealous and punitive. I do not know whether that is strictly true or not, though it is understandable, for a simple man reflecting upon his life and that of his fellows must feel that it is fundamentally a struggle against hostile forces. I have not the effrontery to engage in Biblical exegesis, but would venture to say that when one reads the Book of Job for the first time, one is horrified that an all-wise Father could so try his servant. Yet such trials are horrifying only if one presupposes that life ought to be serene.

But that is precisely what the religious temperament does tempt us into believing. To have faith that everything will come out all right in the end, though the end may not arrive for millennia, that good will triumph and evil be suppressed, is certainly not grounded on anyone's individual experience. In Christian societies people have thought that death, surely the one event that can be counted upon, was a punishment. Man would not have died had not Adam sinned. The Greeks suspected that immortality without youth was not very desirable: witness the myth of Tithonus. Since we could not retain our youth with all its attendant joys, our happiness became *post mortem*. The idea that felicity could be found only beyond the grave was an admission of the sterility of living. No one had the courage to formulate a philosophy of life which would accept death as a conclusive termination, if the phrase is not redundant, and to build up a system of values with the two termini of birth and death. To look forward to utter extinction as the Buddhists did seemed impossible. For that matter it did not take many generations of Buddhists themselves to reject that solution. And the effect of science upon philosophers, as far as this point is concerned, has been to urge them on to new definitions of immortality, such as posthumous fame.

This was not, however, inevitable. It would have been possible for philosophers, following the lead of Aristotle, Spinoza, and the early Bertrand Russell, to construct a religion for a purely terrestrial life. Most of our values would be meaningless in any other life, whether prenatal or *post mortem*. If there are to be no marriages in heaven, then what does the love of men and women mean after death? If we see

God face to face and find our perfect felicity in contemplating the divine essence, then terrestrial truths are not even a step toward that truth, for one cannot know the infinite step by step. The West has seen a steady growth of religious freedom since the sixteenth century, and the result has been an equally steady amendment of some of its more barbarous tenets. Mr. D. P. Walker has recently published a book called *The Decline of Hell* in which he writes the story of the rejection of eternal torture. To define the pleasure of the elect as the sight of their brothers burning and howling in fire and brimstone no longer seems a satisfactory account of beatitude. But that is partly, though not wholly, a result of concentrating our attention on the hells in which we live here and now. One could with a little effort define sin in the context of earthly existence. The old seventeenth-century thesis that a society of atheists could be as virtuous as a society of theists has never been refuted except dogmatically. It hardly could be refuted since examples are there to be observed by all. I may be wrong in thinking that theological dogmas are not to be weighed by their moral effects, but are a matter of faith. But it can be demonstrated that even when faiths vary, brotherly love can still be practiced. There is no country in the world where brotherly love is more widespread than in the United States, though here it is often accompanied by ruthless brutality too. Nor is there any country where faiths are so varied. In fact charity as externally manifested is almost the only religious virtue widely observed here. There is no bond between this virtue and science. But there is a virtue which is inherent in scientific practice that religious societies would do well to emulate.

I refer to the scientists' devotion to truth. This is not only an individual matter but a social matter as well. Whereas statesmen have quarreled with foreign statesmen, scientists have cooperated with foreign scientists. There is no American physics or Russian physics; there is just physics. Moreover, on the national scene, it takes no close scrutiny to see that all science is a social enterprise. The bibliographies that follow scientific studies prove that. I am not neglecting the contributions of individuals to scientific progress, but there is a frank dependence on the work of one's predecessors and contemporaries in all scientific work. I do not say that every scientist is as scrupulous as a saint in admitting his intellectual debts. But the general practice is virtuous, and we can judge only from the general practice. In the humanities the same thing is more or less true, though there descriptive judgments are not so possible and cooperation cannot exist as it does in a laboratory. I doubt very much that members of a church observe the ethical precepts of their creed to the extent that a scientist observes his professional ethics. One reason for this may be that religious people have a feeling that each man must save his own soul in his own way, whereas in science the way to truth is pretty well formalized.

I have said little or nothing of theological questions in this lecture, for I am incompetent to say anything of any importance in that field. I might say, however, as a historian, that theologies based on either metaphysics or science seem futile. The attempts to turn Aristotle's Unmoved Mover into the God of the Bible have all been sterile. The Unmoved Mover was a person only grammatically; he was neither a creator nor a judge of human conduct. You could not pray to him. Inasmuch as he was no more than a cosmic active reason, he could feel neither love nor approbation and hence could not judge men's conduct. In both Christianity and Judaism, God was a being sufficiently like a man to be man's model, and His intervention in human affairs ran from the punishment of sin to reward for purity. It would be absurd to think of the Unmoved Mover in either of these roles, just as it would be sacrilegious to substitute for the Biblical God Hegel's Absolute, Tillich's Ground of Being, or Whitehead's Principle of Concretion, to say nothing of those other metaphysical constructs which are so devoid of concern for human fortunes. It may well be that the Biblical God is not needed as an explanatory hypothesis by science, but whether that is so or not the metaphysical abstractions that are sometimes put in His place are tricky substitutes. My point here is simply that you cannot find the religious God by metaphysics. And I doubt that He can be found by science either. When He has entered into science it has always been as a First Cause which in the long run would never be invoked if the scientist did not want to pay his respects to the Bible.

The very notion of a First Cause is of doubtful coherence. It falls to pieces as soon as it is analyzed. It becomes like the first movement of time or the last point in space, something of more interest to lovers of antinomies than to men who feel awe, helplessness, fear, and trembling when confronted either by Pascal's two infinites or by the insignificance of man as a cosmic detail. Such men will never be swayed by reasoning, for their evidence is that mysterious experience known as faith. It was pointed out centuries ago that there is a mystery at the very heart of existence, and I gather that such men find their religious beliefs to be a solution of the mystery. It can only be said to them that a mystery solved is no longer a mystery. If the philosopher or the scientist thinks that he can put in the place of religious faith his theorems and neatly drawn deductions, then he is self-deluded. He seems to operate on the principle that anything real is rational. But he overlooks experience, which is a blemish on the world of reason. And there is where religion stands firm.

Judaism and Christianity are religions that arose and developed in a world whose limits were the solar system, whose center was the Mediterranean basin. However widespread those limits have become, the immediate human problems still arise on this earth, not on one of the other planets or in the Milky Way or in those regions of space beyond our galaxy. To take the point of view of Sirius will not help solve them; it will merely lead us to neglect them. Birth and death, sickness and health, love and hate, the abuse of power, cruelty, mendacity, ignorance—these and similar things remain our problems. We may no longer pray for the alleviation of suffering, but on the contrary try human means to lessen it. Those means were discovered and improved by applied science. We ask for miracles only when science gives out. But unless I have completely misread Leviticus and Deuteronomy,[7] our religious ancestors were not different from us in this respect. They faced the same problems and also tried to solve them as far as they could, without resorting to super-

[7] For the sake of those who have forgotten their Old Testament, let me refer to Leviticus 19 and 20, and Deuteronomy 5, where the Decalogue is repeated. It is interesting that whereas on the whole we still consider incest, adultery, and sexual perversions reprehensible, we have not retained the punishments for them that are prescribed in these parts of Scripture.

natural aid. Some of these problems have never been solved. I refer to those that seem inherent in the human psyche, those that arise from jealousy, envy, ambition, competition, and all that stimulates the man who has a high degree of the sense of his own importance. But that we call such attitudes vicious comes not from either philosophy or science, except very remotely; it comes from religion. One has to interpret its religious bases allegorically to be sure, for the bald stories in the Old Testament can no longer be taken seriously. The fact, however, that we retain the lessons of such stories and use them as bases of conduct, if not as guides to action, is the hard nucleus which we seldom even try to dissolve. It is with this idea that I close, hoping that I have raised certain problems and suggested ways of answering them. I have no claim to be a religious leader, much less a prophet. But I might venture the opinion that if we ever reach a point in civilization at which there will be peace on earth, it will come about by the inspiration of our religious tradition, elaborated into philosophic principles, and tested by the methods of the scientists.[8]

[8] This is debatable and is given as my own opinion. That opinion rests on what I take to be a fact, namely that our notions of sin and crime come from Scripture, that the reasons we give for considering certain acts as sinful or criminal come from philosophy, and that correction of these reasons comes from science. This seems to me to be historically justified. But should anyone doubt the value of historical fact in philosophic discourse, then of course what I have said is hardly to the point. One might pursue the matter a bit further and

ask whether some of the acts which we find immoral are not repugnant to that basic human nature of which literature is always finding new examples. Basic human nature ought to be really universal and not merely Occidental or north European or Judaeo-Christian. But if my own reading is worth citing, I should say that one finds these universal traits only by the use of common nouns or adjectives which by their very nature eliminate local differences. There is no culture, as far as I know, that disapproves of love. But distinctions are always made between love of other human beings and love of God; between love between man and woman and love between two people of the same sex; between love for a woman not consanguineous to some degree or other and love for a woman whose relationship falls within the prohibited degree. In short the basic drives, as they are sometimes called, cannot be annihilated by fiat, but every society regulates their satisfaction and does so in its own way.

THE CHALLENGE TO RELIGION

George Boas, Professor Emeritus of the History of Philosophy at the Johns Hopkins University, has published seven books since his official retirement in 1957. He has also continued to teach, having served as Mellon Professor of Philosophy at the University of Pittsburgh in 1960-61 and as a Fellow of the Center for Advanced Studies at Wesleyan University in 1961-62.

Born in 1891, Boas had earned Bachelor and Master of Arts degrees from Brown University, as well as a Master of Arts from Harvard by 1915. Two years later he acquired a Doctor of Philosophy degree from the University of California. After two years' service as an army lieutenant in the First World War, he returned to teach at the University of California from 1920 until he joined the faculty of the Johns Hopkins University in 1921. His association with that university has continued to the present, interrupted only by service as a commander in the navy during the Second World War.

In his books, Boas deals principally with the history of philosophy and with aesthetics. His most recent publications include Wingless Pegasus (1950), Dominant Themes of Modern Philosophy (1957), The Inquiring Mind (1959), Some Assumptions of Aristotle (1959), The Limits of Reason (1961), Rationalism in Greek Philosophy (1961), The Heaven of Invention (1963), and What Is a Picture? (1964, with H. H. Wrenn).

Professor Boas has been awarded honorary degrees from the Johns Hopkins University, Washington and Lee University, the University of New Mexico, and Washington College. He holds membership in many professional organizations, including the American Philosophical Society, the American Philosophical Association, the American Society for Aesthetics, and the American Academy of Arts and Sciences, and is also an Associé of the Académie Royale de Belgique.